Social Media and Interactive Communications

The extent to which social media can potentially add value within various service contexts is not well understood. While at a general level it would seem that direct and immediate interactive communication with customers and stakeholders would be of benefit in terms of general communications, the integration of new media alongside more traditional marketing activities is not without difficulty. Many organisations appear seduced by what new technological communication channels are capable of but evidence suggests that those same organisations may have limited sensitivity to the appropriateness of employing social media to add value to the customers' service experience. Launching social media initiatives appears low cost and fairly straightforward, technically, but managing the subsequent interactions and engagement appropriately, and indeed profitably, can often be beyond a firm's resources and competencies. In this book the challenges of effectively managing interactive communications through social media is described in various service contexts (e.g. healthcare, travel, small businesses) and within prevailing, yet ever more crucial marketing concepts, such as customer relationship management (CRM) and customer complaining behaviour.

This book was originally published as a special issue of *The Service Industries Journal*.

Mark Durkin is Professor of Marketing in the Ulster University Business School, UK. Professor Durkin spent over a decade in various service, sales and marketing roles with the Bank of Ireland Group, culminating in the role of Group Marketing Communications Executive (NI). He has worked at Monash University in Melbourne, Australia, and Otago University in Dunedin, New Zealand, and in 2006 was awarded the Chartered Institute of Marketing's (CIM) 'Marketing Person of the Year' in recognition of his contribution to marketing education in Ireland. He has conducted his research and presented to academic and practitioner audiences in Australia, New Zealand, Sweden, USA, UK and Ireland, and he is currently focused on exploring the growing domain of digital marketing. He works closely with business practitioners in helping improve business practices especially as these relate to the adoption and utilisation of new technology within marketing.

Aodheen McCartan is Senior Lecturer in the School of Communication at Ulster University, UK. Her research interests lie in the fields of marketing communication, small firm marketing and financial services marketing. A focus of her research is how technology can complement traditional communication platforms.

Mairead Brady is Assistant Professor of Marketing at Trinity College Dublin, Ireland, and is also the co-author of the popular *Marketing Management* textbook with Philip Kotler, Kevin Keller, Malcolm Goodman and Torben Hasen. Her research interests and publications are in the areas of technology and marketing practice, social media, data analytics and gamification.

Social Media and Interactive Communications

A service sector reflective on the challenges for practice and theory

Edited by
Mark Durkin, Aodheen McCartan and Mairead Brady

Routledge
Taylor & Francis Group

LONDON AND NEW YORK

First published 2017
by Routledge
2 Park Square, Milton Park, Abingdon, Oxfordshire OX14 4RN
711 Third Avenue, New York, NY 10017

Routledge is an imprint of the Taylor & Francis Group, an informa business

First issued in paperback 2018

British Library Cataloguing in Publication Data
A catalogue record for this book is available from the British Library

ISBN 13: 978-1-138-65859-2 (hbk)
ISBN 13: 978-0-367-02396-6 (pbk)

Typeset in Times New Roman
by RefineCatch Limited, Bungay, Suffolk

Publisher's Note
The publisher accepts responsibility for any inconsistencies that may have
arisen during the conversion of this book from journal articles to book chapters,
namely the possible inclusion of journal terminology.

Disclaimer
Every effort has been made to contact copyright holders for their permission to
reprint material in this book. The publishers would be grateful to hear from any
copyright holder who is not here acknowledged and will undertake to rectify
any errors or omissions in future editions of this book.

Contents

Citation Information

The chapters in this book were originally published in *The Service Industries Journal*, volume 35, issues 11–12 (August 2015). When citing this material, please use the original page numbering for each article, as follows:

Chapter 1
Social media's contribution to customer satisfaction with services
Mary Beth Pinto
The Service Industries Journal, volume 35, issues 11–12 (August 2015), pp. 573–590

Chapter 2
Extending customer relationship management into a social context
Sarah Diffley and Patrick McCole
The Service Industries Journal, volume 35, issues 11–12 (August 2015), pp. 591–610

Chapter 3
The value of social presence in mobile communications
Ji Hee Song and Candice R. Hollenbeck
The Service Industries Journal, volume 35, issues 11–12 (August 2015), pp. 611–632

Chapter 4
Customer e-complaining behaviours using social media
M. S. Balaji, Subhash Jha and Marla B. Royne
The Service Industries Journal, volume 35, issues 11–12 (August 2015), pp. 633–654

Chapter 5
Marketing technology for adoption by small business
Philip Alford and Stephen John Page
The Service Industries Journal, volume 35, issues 11–12 (August 2015), pp. 655–669

Chapter 6
Exploring interactive communication using social media
Chrystal B. Zhang and Yi Hsin Lin
The Service Industries Journal, volume 35, issues 11–12 (August 2015), pp. 670–693

For any permission-related enquiries please visit:
http://www.tandfonline.com/page/help/permissions

Notes on Contributors

Philip Alford is a small business marketing specialist in the Faculty of Management, Bournemouth University, Fern Barrow, UK. His research interests lie particularly in online marketing technology. He has published widely on the subject, in journals such as, the *British Journal of Educational Technology*, *Technovation*, and *Travel & Tourism Analyst*.

M. S. Balaji is Lecturer in the Taylor's Business School, Taylor's University, Selangor, Malaysia. He examines service failure and recovery, relationship marketing and social media. Dr Balaji was the recipient of the AIMS International Young Management Researcher Award in 2012.

Mairead Brady is Assistant Professor of Marketing at Trinity College Dublin, Ireland, and is also the co-author of the popular *Marketing Management* textbook with Philip Kotler, Kevin Keller, Malcolm Goodman and Torben Hasen. Her research interests and publications are in the areas of technology and marketing practice, social media, data analytics and gamification.

Sarah Diffley is a Lecturer in Marketing at Letterkenny Institute of Technology, Co. Donegal, Ireland. She received her PhD from Queen's University Belfast. This study is based on her PhD research. Her main research interests include empirical research in customer engagement and value co-creation, the role of social media technologies in customer engagement and value co-creation, and many-to-many marketing communications.

Mark Durkin is Professor of Marketing in the Ulster University Business School, UK. Professor Durkin spent over a decade in various service, sales and marketing roles with the Bank of Ireland Group, culminating in the role of Group Marketing Communications Executive (NI). He has worked at Monash University in Melbourne, Australia, and Otago University in Dunedin, New Zealand, and in 2006 was awarded the Chartered Institute of Marketing's (CIM) 'Marketing Person of the Year' in recognition of his contribution to marketing education in Ireland. He has conducted his research and presented to academic and practitioner audiences in Australia, New Zealand, Sweden, USA, UK and Ireland, and he is currently focused on exploring the growing domain of digital marketing. He works closely with business practitioners in helping improve business practices especially as these relate to the adoption and utilisation of new technology within marketing.

Candice R. Hollenbeck is Lecturer in the Department of Marketing, Terry College of Business, University of Georgia, Athens, Georgia, USA. Her work examines the socio-cultural and symbolic dimensions of consumption and the cultural ecology of marketing. Her recent articles were published in *Business Research*, the *International Journal of Research in Marketing*, and *The Service Industries Journal*.

Subhash Jha is Assistant Professor in the Department of Marketing, Indian Institute of Management, Udaipur, India. He specializes in haptic information processing, the service loyalty model and advertised reference pricing. He has had articles published in *Marketing Intelligence and Planning*, the *Journal of Consumer Marketing* and the *Asia Pacific Journal of Marketing and Logistics*.

Yi Hsin Lin is Associate Professor in Business Administration and Marketing in the Department of Leisure and Recreation Management, Asia University, Taichung City, Taiwan.

Aodheen McCartan is Senior Lecturer in the School of Communication at Ulster University, UK. Her research interests lie in the fields of marketing communication, small firm marketing and financial services marketing. A focus of her research is how technology can complement traditional communication platforms.

Patrick McCole is Senior Lecturer in Management in the Queen's University Management School, Queen's University Belfast, Belfast, Northern Ireland. His research interests include online purchasing behaviours, crowdfunding and e-business. His recent publications have appeared in the *Australasian Marketing Journal*, *Marketing Intelligence and Planning* and the *Journal of Business Research*.

Stephen John Page is Professor and Deputy Dean in the Faculty of Management, Bournemouth University, Fern Barrow, UK. He specializes in tourism management. His latest project looks at sustainable tourism policy and practice and the role of tourism and seasonality. He is co-author of *Event Studies: Theory, Research and Policy for Planned Events* (Routledge, 2015).

Mary Beth Pinto is Professor of Marketing in the Black School of Business, The Behrend College, Pennsylvania State Erie, Erie, USA. Her research interests include consumer behaviour, services marketing, social media, pedagogy and e-learning. Professor Pinto's work has appeared in such journals as the *International Journal of Project Management*, *Journal of Higher Education Theory and Practice* and the *Journal of Management and Marketing Research*.

Marla B. Royne is the Great Oaks Foundation Professor of Marketing in the Department of Marketing and Supply Chain Management, Fogelman College of Business and Economics, The University of Memphis, Memphis, USA. Her recent publications can be found in the *Journal of the American Academy of Child and Adolescent Psychiatry*, *Journal of Consumer Affairs* and the *Journal of Business Research*.

Ji Hee Song is a Researcher in the College of Business Administration, University of Seoul, Seoul, South Korea. His research focuses on human-computer interaction and information systems (business informatics).

Chrystal B. Zhang is Senior Lecturer in the Faculty of Science, Engineering and Technology, Swinburne University of Technology, Melbourne, Australia. She specializes in aviation and marketing, and has published in such journals as the *Journal of Travel and Tourism Marketing* and the *Journal of Air Transport Management*.

INTRODUCTION

Enabling interactive communications through social media: Research from services contexts

Mark Durkin,[a] Aodheen McCartan[b] and Mairead Brady[c]

[a]Ulster University Business School, Ulster University, UK; [b]School of Communication, Ulster University, UK; [c]Trinity College Dublin, Dublin, Ireland

When a technology like social media becomes so pervasive in society, business must be on alert to the consequent changes for their customers. When technology more broadly changes the balance of power of customer relationships, then businesses need to be more than alert – they need to be at the forefront of the challenges and opportunities which emerge and they must be positioned effectively to manage these challenges and capitalize on the opportunities. This book supports businesses generally and service businesses in particular, which are seeking to engage, with what can seem like an ever changing digital phenomenon. We explore many of the managerial and behavioral changes which will continue to alter the business domain. This book provides managerial insight into a range of social media manifestations from e-complaining, to empathy in text messages, from the need for speed in response times, to adoption challenges. All provide managerial insight and support for service businesses grappling with social media and digital challenges.

We include the word 'enabling' in the title of this introduction to suggest a positive focus towards action and empowerment and towards patterns of interactions which allow for development and growth. To enable companies to really engage with social media we need to provide them with the support that is not yet there, with education that has not developed fast enough in order to guide them in how to work and succeed in an environment that is always changing and at speeds never witnessed before. In a world charged with technological change this book studies how businesses could and should engage with customer communication through social media. What service businesses need is new models of interactive communications which bridge the gap between theory and practice and in so doing better address both the challenges of social media as well as framing the opportunities in a more accessible way.

There is a sense of social media compelling companies to engage with 'always on' marketing programmes (Von Bummell et al., 2014), representing a shift from how business has traditionally been practised. Consumers have changed with mobile abounding (Shcadler, Bernoff & Ask, 2014), with the always on consumer (Joachimsthaler, 2014; Turkle, 2012), and the internet in the pocket generation (Bell, 2015) so the consumer is always connected. But a key questions remains – how can service companies leverage this new platform for commercial advantage and use it to potentially add value to their customer relationships? We suggest that as social media has become more critical in the consumer decision journey it is increasingly used by customers for purchase decision support. Therefore, businesses need to have the competencies to capitalize on that opportunity: to have content, that is, to have something engaging to say over time and to deliver such content in a way that is context-specific for the customers of that business (Durkin, 2013). This book is focused on the axis between theory and practice in service contexts. The book is about enabling and empowering companies in the social media era through theoretical ideas and concepts alongside practical guidelines for growth and development.

Each of the chapters in this book draws on developing literature in this area. It provides an alignment of the findings within both the literature and practice. Readers can therefore explore both the theoretical discussions juxtaposed with what is happening in practice. These chapters are a support for service businesses to give them the insight they need but they are also useful as a broad overview of this developing but embryonic literature.

The chapters individually and collectively build to a depth of exploration of a variety of academic theories and often link theories not previously aligned. Chapter 1 provides depth in the theoretical arena of social capital theory and patient centred theory while chapter 2 draws on customer relationship management, the resourced based view, value co-creation, service dominant logic and relational information processing. Chapter 3 explores social presence theory and interactive theory while chapter 4 explores the use of justice perception theory, attribution theory and interestingly aligns it with self-categorization theory, an area which has developed so much with the concept of the self and the digital self. The last chapter introduces functional and contingent interactivity. All of these provide a wealth of discussion and interesting perspectives which allows for a real engagement with these issues, which is helpful for not only academics but also for practitioners in services marketing contexts.

The six chapters cover a broad range of the research and practice issues service businesses are encountering including ePatients, e-complaining, text message content, technology adoption and the speed and focus of responses. Each contribution was chosen to demonstrate the breadth and depth of the social media challenges and opportunities, and because it made a valuable and interesting empirical or conceptual contribution to advancing knowledge in this area. The contributions are truly international, with authors from Great Britain, Northern Ireland, America, Australia, Korea, Malaysia, India and Taiwan.

The first chapter is from Mary Beth Pinto from the USA and here, social media's contribution to customer satisfaction within the health services context is examined. This chapter explores how consumers (patients) can take control through an ePatient model to guide and direct their own service provision. Managing service delivery at the consumer interface, when social media is now present, changes the traditional patient doctor dyadic relationship. This chapter takes the patient perspective within a patient centred managed (PCM) model and studies firm hosted sites. What was explored was how the internet gave the patient autonomy. They suggest that 'what we are beginning to see is a convergence of multiple patterns of social and technology driven behaviour that have powerful implications for the future of health care service delivery' (17). More and more patients have information to hand through social media, and the health services, like other sevices, can capitalize on this, though issues of boundary and confidentiality might of course impact upon this. The findings were very positive from the patient perspective with social media seen as an excellent support for the patient centred management practice, with interview evidence that the patients really enjoyed the level of information, connection and support they received through social media.

More interesting, but worrying from a managerial perspective, was the lack of resources and expertise within the hospital research site to develop social media based content on a regular basis. The study found that they were relying on a local university to provide this on an ad hoc basis. They showed how, though there was a lack of engagement with social media, the hospital did see themselves as a social community and enjoyed the speed and efficiency needed for creating and transferring knowledge through social media.

For businesses, what is needed but may not be there includes:

Company resources – Continuous time, people and money for the employment of staff, expert in ongoing social media to create on an ongoing basis, new, fresh and engaging content;
Patient/customer resources – A patient panel that understands and could drive the development of content;
A commitment to **refashion traditional** roles including the role of doctor and the role of patient;
A commitment to **redefine the structure** of social networks.

This chapter confirms the support and satisfaction that customers, in this case patients, received through social media but also showed how the 'supplier' was not ready to fund or develop this as a core aspect of its 'business'.

The second chapter also discusses the use of the information to inform and develop the relationship with the customer. This chapter by Diffley and McCole is entitled 'Extending customer relationship management into a social context' and focuses on the hotel sector, exploring what social capabilities and what depth of network interactions are necessary for co-creation of value. What the authors suggest is that information and interaction management represent key capabilities that should be fostered. It is further suggested that the direct and positive impact of relational information processing on customer and financial preformance demonstrates the benefits of integrating social media into organizational routines and that management must acknowledge the importance of social networking sites in enhancing relationships, and thus integrate these sites into their businesses. This chapter takes a very positive perspective to how social media can enhance the customer relationship and provides a helpful critique of customer relationship management (CRM) for any companies grappling with the concept. The authors suggest that IT competencies within the organization and a social media orientation are both imperative. They conclude that social media has revolutionized and perceptibly improved CRM by allowing true co-creation to occur. Any company struggling with CRM will find this a welcome perspective which might finally show them how to really engage with customers and also what human, technical and business resources are needed to do so.

The third chapter entitled 'The value of social presence in mobile communications', written by Song and Hollenbeck, focuses again on aspects of management skill-set but this time on mobile technology and text messages via social media platforms. The authors suggest that companies need to be able to reply quickly in a concise manner, but are concerned that there is a lack of skills in this domain. They use social presence theory to suggest that human warmth and empathy must be conveyed even within short text messages. Businesses, generally, have developed necessary face to face communication skills, such as verbal cues, body language and emotional displays, but there are relatively fewer companies that have learnt how to incorporate social presence cues into texting. Perceived interactivity has three features – perceived two way communication, perceived control and perceived responsiveness, and the researchers linked these to social presence cues. Social presence theory proposes that adding human warmth positively affects interactivity perceptions and subsequently enhances customer satisfaction, attitude and repurchase intention. This research shows that greater verbal social presence cues are associated with greater levels of satisfaction and that customers express a preference for social presence cues particularly empathy and intimacy from companies.

Social media has allowed promotion to move beyond one way communication towards two way communication. This paper underscores that even in short abbreviated sentences we can still instill warmth and empathy. It is suggested that companies need a planned process for service level recovery or intervention strategies and that text messages should be concise but imbued with social cues, it is noted that even with appropriate levels of social cues in the text, this will not compensate for unsuccessful service recovery. Negative impressions can go viral fast and so business reactions must be designed to counteract and manage that by responding quickly and appropriately. Mobile texting is a viable means to handle service complaints and can help to change or improve the virtual mood. The authors call for social presence cues to be considered for all virtual communications as a means of increasing customer satisfaction and attitude to the firm. The authors conclude by stating that customers value time and convenience but they also value warmth!

The fourth chapter also focuses on e-complaining and is titled 'Customer e-complaining behaviours using social media' and is written by Balaji, Jha and Royne. They study a real concern of businesses, that of public complaining (to service providers, but specifically over a social media platform like Facebook) and private complaining (to other customers hidden from company view). They suggest web care interventions and note that unfairness, firm response, locus of issues and personal identity are all strong influences on public e-complaining while the desire for retaliation is a major factor in private complaining. The ease of complaining online makes complaining more likely and there is a customer expectation of a speedy reply. This chapter studies what motivates customers to complain using psychological theories (the fairness preception theory, attribution theory and self-categorization theory) and also propose web care interventions and the situational or personality factors which could determine whether the customer is more likely to engage in public or private complaining. If the customer is complaining using the company's social media platform, that is, a form of public complaining, then the company needs to be alert and have the necessary financial and human resource investment in place, as speed is of the essence. They suggest that companies ought to acknowledge, recognize and seek to recover while all the time being aware that other customers are watching, who might also need to know the solution. For private complaining they advise a rigorous system to track the social media complaints, which includes trying to handle them on the same medium and ensuring to point out that the company was not aware of the complaint but is now wishing to resolve it.

Chapter five, by Alford and Page, takes the perspective of the adoption of marketing technology by small firms and focuses on the challenges for small owner managed firms as they struggle to reap potential advantages. They note that often there is a lack of knowledge, an inability to measure return on investment, a lack of a clear eVision and a lack of clear perception of benefits. The study found that the dominant challenges were what most owner-managers would already be familiar with including, market orientation, strategy and vision, barriers, measurement, collaboration and benefits but that despite these issues there was an overwhelmingly positive attitude and a real desire to adopt the technology. They suggest that what is needed is a move from a business centric to a customer centred view of the business. The key challenge is to turn the attitude into action and in so doing, develop the new skill sets needed to analyze data, measure ROI, integrate customer touchpoints and create engaging content within the context of the culture of the organization. They suggest that firms find it difficult to learn and that there needs to be a greater adoption of a test and learn mentality. They provide a model with three stages, as follows:

1. Gain insights of target market;
2. Focus on consumer journey and acquisition and retention;
3. Measure, test and learn in a feedback loop.

What adds value in this study is that they then design a model for the owner in how to gain knowledge, understand the benefits, explore the resource constraints and align marketing technology with the consumer decision journey.

Chapter 6, 'Exploring interactive communication using social media' is the concluding chapter in the book. Written by Zhang and Lin, it focuses on level and type of interactivity. What the authors suggest is that there is evidence of functional interactivity, from a technology perspective, but that individuals are focusing on contingent interactivity. Functional interactivity is a dialogue or informational exchange between the user and the interface and includes such features as, an email link, chat room facility, survey on the site, polls and any format which says: we want to hear from you. They are essential in facilitating a dialogue loop which engages customers online. Contingency interactivity is a process involving users, media and messages in which the communication roles are interchangeable and interactive to customers. The messages in the interactive process are contingent on the previous messages and the responses are intertwined and cumulative. Using the interactive performance matrix there is evidence that some companies are adopting a customer centric approach but there is also evidence that other companies are struggling in this domain. The authors suggest how companies can use the interactive features to support relationship development. The consumer holds the power and real time conversations without time and physical constraints can be very difficult for companies to really understand as well as the possibility of not having the systems in place which allow for interactivity.

The authors ask if companies are ready for contingent interactivity and whether they have in place the systems designed for this. In the past there might have been an urge to closedown the interaction as quickly as possible and to hope that the customer will be satisfied with the speed at which the complaint was handled. This chapter reminds us all that many consumers just don't care about companies; they are on social media to entertain themselves, to pass the time, to gather information, as a means of self expression and self satisfaction. The authors conclude that customers perceive more added value in a more personalised interaction, but businesses prefer an automated interaction. Using airlines as an example, they note that although KLM posted the most on Facebook, they were still outposted by customers with 1,800 posts by KLM and 2,900 by customers in a week. It is interesting to note that Ryanair made no posts which highlights a difference in a full service and low service model. Though full service accounted for 60% of airline initiated efforts, the dominant focus was on advertising and sales promotion. The real challenge lies in the difference between what the customer is posting, which is predominantly information seeking and e-complaining, compared to the content of the marketer generated posts. Of the 20 airlines mentioned it was surprising to note that 17 received more grievance-type posts than compliments except for three: KLM, Swiss Airlines and Southwest Airlines. What the authors suggest is that instant reach reduces stress for airline passengers. They provide a matrix of profiles including leaders, user activist, business activist and laggards. The laggards are mainly the LCC. Their managerial suggestion is that businesses need to consider the type of messages they send, and that the current style of promotion and advertising will not generate the type of engagement needed.

To conclude, all the chapters explore the managerial challenges associated with a new world of social media and digital engagement. The book showcases services industries that are engaging well with the digital challenges but it also showcases companies that are still struggling both with the technology itself but mainly with being limited in the resources of managerial time, knowledge and appropriate staff. The book explores the commitment needed across organizations if they wish to fully engage and become 'digital businesses'. We hope that you enjoy this book which is packed full of great stories, instructive case studies and practical ideas on how to use social media within both small and large service organizations. It provides valuables insights into real practices and processes that can, with the investment of time, a customer-orientation and appropriate resources, truly reap great benefits.

References

Bell, David. (2015). 'Millennials on Steroids': Is Your Brand Ready for Generation Z? *Wharton*. http://knowledge.wharton.upenn.edu/article/millennials-on-steroids-is-your-brand-ready-for-generation-z/

Durkin, M. (2013). Tweet me cruel: perspectives on battling digital marketing myopia. *The Marketing Review*, *13*(1), 51–63.

Joachimsthaler, Erich. (2014). Divining the Future, The Always on Consumer, *Forbes.* http://www.forbes.com/sites/onmarketing/2014/02/27/divining-the-future-the-always-on-consumer/#1dd6d6581b18

Shcadler, Ted, Bernoff, Josh, & Ask, Julie. (2014). *The Mobile Mind Shift: Engineer Your Business to Win in the Mobile Moment*. Groundswell Press.

Turkle, Sherry. (2012). *Alone Together: Why We Expect More from Technology and Less from Each Other*, first trade paper edition. New York: Basic Books.

Von Bommel, Edwin, Edelman, David, & Ungerma, Kelly. (2014). Digitizing the consumer decision journey. *McKinsey*, June, 1–8.

Social media's contribution to customer satisfaction with services

Mary Beth Pinto

Penn State Erie, The Behrend College, Black School of Business, Erie, PA, USA

An important recent initiative in the effective transmission of healthcare services is the establishment of the patient-centered medicine (PCM) philosophy as a mechanism for enhancing customer satisfaction. Although the goals of PCM are important, there is less understanding of the means by which service providers can promote this philosophy.

This study examines the relationship between customers' attitude toward and use of social media, PCM, and their satisfaction with healthcare services. Data were collected from a large, urban-based pediatric office in the northeast. The sample consisted of 234 respondents who were classified as 'e-Patients' – that is, they reported having access to the Internet and going online for health information. A three-stage regression analysis, conducted to establish the path coefficients for each stage in the model, shows that customers' (patients') attitude toward social media can be an effective method to enhance PCM and, ultimately, satisfaction. The findings contribute to theory in services by exploring the challenges of managing service delivery at the interface between customer satisfaction and the role and usefulness of adopting and effectively using social media.

Introduction

Healthcare service in the USA continues to be in a state of dynamic metamorphosis. Over the past 40 years, a number of initiatives, some public policy-based and others resulting from changes in technology and population demographics, have led to rethinking healthcare service delivery and the manner in which professionals are expected to practice. The recent introduction of the Patient Protection and Affordable Care Act in March of 2010 and the resulting controversies it has created present challenges for health service professionals, as they come to terms with the overriding goal of providing optimal patient care within a complex healthcare environment.

A critical paradigm for the nation's healthcare delivery has been the adoption and expansion of patient-centered medicine (PCM). Interest in the topic dates back to the late 1960s and has continued over the last 50 plus years (Balint, 1969; Byrne & Long, 1976). The trend toward patient-centered medical care accelerated in 2001 when The Institute of Medicine identified PCM as one of the six goals for healthcare delivery in the USA. Specifically, PCM was identified as 'care that is respectful of and responsive to individual patient preferences, needs, and values' (Institute of Medicine, 2001, p. 3).

Since that time there have been numerous conceptualizations of PCM. Most frameworks agree that patient-centered decision making is 'the process of identifying clinically relevant, patient-specific circumstances and behaviors to formulate a contextually appropriate care plan' (Weiner et al., 2013, p. 573). Within PCM, effective care is seen to be collaborative – that is, defined by or in consultation with patients rather than by physician-dependent tools or standards.

The PCM initiative is rooted in Social Capital theory, as articulated by sociologists and organizational researchers over the past two decades (cf. Adler & Kwon, 2002; Anheier, Gerhards, & Romo, 1995; Lin, Cook, & Burt, 2008). Social capital generally refers to 'the sum of resources, actual or virtual, that accrue to an individual or group by virtue of possessing a durable network of more or less institutionalized relationships of mutual acquaintance and recognition' (Bourdieu & Wacquant, 1992, p. 14). It has been instrumental in furthering our knowledge of social interactions within communities, public health, organizational governance, and other problems of collective action. Further, social capital has been associated with several positive social outcomes such as better public health, lower crime rates, and more efficient financial markets (Adler & Kwon, 2002; Szreter & Woolcock, 2004). A recent example was the decision by the Cleveland Clinic to ask every patient calling in to seek an appointment if they would like to be seen that same day. Their reasoning was based on the principle of nurturing a collective sense of 'we,' in which all members of the social network partner to improve healthcare quality delivery (Norrish, Biller-Andorno, Ryan, & Lee, 2013).

The central tenet of social capital theory is that it allows people to 'draw on resources from other members of the networks to which he or she belongs' (Ellison, Steinfield, & Lampe, 2007, p. 1145). Social capital researchers believe that Internet-based relationships, such as those created through social media, allow users to develop and maintain larger networks from which they can potentially draw resources and benefits through the exchange of useful information, development of personal relationships, and the formation of groups for support or action (Donath & Boyd, 2004; Mathwick, Wiertz, & de Ruyter, 2008; Steinfield, Ellison, & Lampe, 2008). In a recent study on how technology – including the Internet and mobile phones – interfaces with medical care, researchers observed the creation of social capital for pregnant women seeking information and advice on prenatal care (Kraschnewski et al., 2014). The women in the study reported that information obtained from the Internet helped them cope with the health conditions surrounding their pregnancy. The authors concluded 'given how critical patient-provider communication is to the therapeutic relationship, the Internet should be considered by more providers as a forum for both dissemination of evidence-based education information and integration into the prenatal care structure' (Kraschnewski et al., 2014, p. e147). Thus, Social Capital theory supports attempts to pursue a broad patient satisfaction mandate (cf. Morrow, 1999) and, more specifically, informs and shapes initiatives such as the PCM movement herein described.

PCM operates with some important tenets. First, the medical encounter must be viewed and evaluated through the eyes of the patient, not the service provider (Epstein, Laine, Farber, Nelson, & Davidoff, 1996). As a result, it is incumbent upon doctors to recognize the need for the critical skills of empathy and communication in order to better understand patients' perspective and engage with them as partners in their treatment. In their study of several hundred patients, Kim, Kaplowitz, and Johnson (2004) concluded that patient-perceived physician empathy was correlated with a perception of physician expertise, trust, and information exchange, and that such empathy was associated with greater patient satisfaction and compliance. Hojat, Louis, Maio, and Gonnella (2013) concur that empathic engagement in patient care lays the foundation for a trusting relationship and leads to improved patient outcomes.

Second, patients seek, fundamentally, relationships not transactions. The goal of health services should be long-term and empowering partnerships, with each party equally vested in both immediate treatments but more importantly long-term relationships. Little et al. (2001) demonstrated that a personal relationship between patient and doctor and a feeling of partnership led to patients who were more satisfied, more enabled, and had a lower symptom burden and lower rates of referral. Saultz and Lochner (2005) found an association between patients who generally see the same doctor and better outcomes, better preventive care, and fewer hospitalizations. Indeed, the 'therapeutic alliance' paradigm is predicated on developing and maintaining constructive long-term doctor–patient relationships that offer benefits to patients and service providers alike (Mead & Bower, 2000).

Lastly, PCM requires that systems be in place to continually monitor and measure patient perceptions. In other words, it is not sufficient to assume that all is well, that patients are satisfied, and that convenient surrogates (e.g. number of phone calls requesting clarification per month) are useful determinants of satisfaction. Effective PCM efforts also include feedback loops and some form of assessment tools that can be continually accessed by patients and providers alike, to ensure that all parties are satisfied. Therefore, a critical step in understanding patient-centered care is recognizing that patients must be asked to rate or judge their healthcare.

Research suggests that PCM offers a number of advantages for health service professionals, their institutions, and the patients themselves; that is, patient-centered encounters, focusing on the collaborative relationship result in higher patient and physician satisfaction levels, fewer malpractice complaints, while not significantly lengthening the duration of office visits (Stewart et al., 2000). In short, PCM appears to provide both the psychological benefits of enhanced, partnership-based care while also helping medical institutions to keep a lid on costs. It is a low-tech, humanistic approach to medicine that counteracts the assumptions driving a clinical belief in more tests, more technology, and more medication as the key to a successful patient encounter. In fact, so successful has this philosophy been in recent years that a number of medical schools have developed courses in PCM; for example, in 2006, The Stritch School of Medicine at Loyola Chicago incorporated a PCM course into its first- and second-year curricula. It has been a value added component of their curriculum ever since.

Enhancing patient-centeredness: the role of social media

Evidence shows that when patients perceive their medical practitioner has adopted a patient-centered mindset, the effective delivery of healthcare services is enhanced. Less well understood, however, are the best communication tools and technologies to build patient–physician relationships, permitting more fully engaged patients and their families to function as partners in the delivery of their medical care. One critical new means being explored is the use of social media technologies for transforming 'how clinical practitioners, patients and their families work together' (Bacigalupe, 2011). Rozenblum and Bates (2013) suggest, 'healthcare, social media and the Internet – are beginning to come together … and have the potential to create a major shift in how patients and health care organisations connect' (p. 183).

Social media (or technically labeled 'Web 2.0') are 'online means of communication, conveyance, collaboration, and cultivation among interconnected and interdependent networks of people, communities, and organizations enhanced by technological capabilities and mobility' (Solomon, 2013, p. 18). Some social media tools include: social networking applications (e.g. Facebook and Twitter), social collaboration tools (e.g. wikis and blogs), microblogging (e.g. Yammer), and content tracking tools (e.g. Delicious) (Granitz & Koernig, 2011).

These web-based and mobile technologies have become an important part of mainstream society. The Pew Research Internet Project reports that in 2013, 85% of US adults (18 years of age and older) were Internet users (Zickuhr, 2013). According to the Social Media Report (Nielsen, 2011), 'social networks and blogs reached nearly 80% of active U.S. Internet users and represent the majority of Americans' time online' (p. 1).

The advantages of using social media in business are heavily documented (Safko, 2010). Several of the benefits include customer relationship management, customer engagement, branding opportunities, and market intelligence. Over the last several years, increased attention has been focused on how to apply social media to the health services arena. Lane (2010) advocates for the use of social media, the Internet, and mobile-based tools for distributing and sharing information between healthcare providers and patients. She posits that although the healthcare industry has been slow to adapt to social technology changes that social media has 'the potential to educate, engage, and empower patients' (p. 6). In fact, Thielst (2011) notes that the 'ubiquitous nature of social media creates opportunities for true patient-centered care.'

Graham (2011) offers three uses for social media in the health services field: (1) Community development (both professional and personal); (2) Marketing tool for health service providers to promote word-of-mouth communication; and (3) Knowledge dissemination for medical treatments and disease prevention. Pursuing these goals, various social media technologies are now being introduced in the health services to disseminate health information and engage patients with their care (Rozenblum & Bates, 2013). For example, social media technologies often seen in association with medicine and oncology include Blogging, multi-media sharing with YouTube, and Social networking with Twitter, Facebook, and LinkedIn (Graham, 2011). In general, patients are using the Internet and social media to 'connect with others having similar illnesses, to share experiences, and begin managing their illnesses by leveraging these technologies' (Rozenblum & Bates, 2013, p. 183). Research by McDaniel, Coyne, and Holmes (2011) shows that blogging plays an important role in social support. In their study of new mothers and Internet usage, blogging and social networking were shown to improve perceptions of social support, increasing maternal well-being.

It is clear that PCM represents a method for improving the patient–doctor relationship, increasing patient satisfaction with healthcare delivery, while also actively engaging the patient population in their own healthcare. More recently, however, practitioners and researchers have begun to explore how the use of the Internet and social media technologies will help facilitate PCM, as well as investigating patients' receptivity toward and use of social media for healthcare. Thus, our research model suggests the following relationships (see Figure 1): that is, we argue that both the patients' attitude toward the social media and their actual usage of the Internet should be predictors of higher levels of PCM, which, in turn, will improve patient satisfaction. 'Attitude toward the social media' is defined as a person's general evaluation of social media (e.g. how an individual *feels* about social media). The model presupposes that Internet usage and attitude toward social media can have both direct and indirect effects on patient satisfaction, through the mediating role of PCM.

The purpose of this paper is to report on the results of a study aimed at empirically investigating the relationship between patients' attitude toward and use of social media,

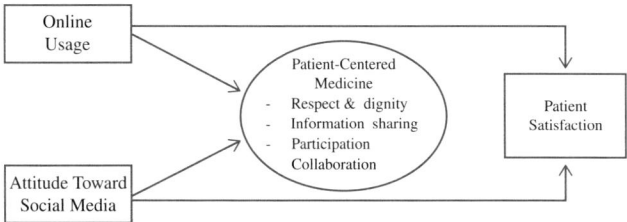

Figure 1. Conceptual model for study.

their perception of a patient-centered approach to the practice of medicine by their primary care practice, and subsequent patient satisfaction. Bacigalupe (2011) has noted that, to date, little empirical research on social technologies has studied its impact on collaborative health, particularly from the perspective of the recipient of healthcare – the patient. Rozenblum and Bates (2013) recognize the need for in-depth studies related to the impact of social media on healthcare consumers and patients (from a social capital theory perspective, Szreter and Woolcock (2004) labeled these impacts serving as 'bridging,' 'bonding,' and 'linking' perspectives). With the rapid increase in the types of social media available for healthcare consumers and the interest in the use of additional, electronic means for pursuing collaborative healthcare, it is critical that we begin to determine how social media can be used to enhance PCM.

The shape of 'network ties' within a social media network is a critical feature of social capital enhancement (Adler & Kwon, 2002) as it defines not only the flow of opportunities, motivation, and ability of influence of social capital quality but the structure of these networks. In this sense, social media represents a network content element useful for permitting patients to create and enhance social ties and improve social capital relationships, including those with their medical practitioners. Recalling Kogut and Zander's (1996, p. 503) contention that 'a firm be understood as a social community specializing in the speed and efficiency in the creation and transfer of knowledge,' we can examine the role of social media as a mechanism for this knowledge transfer and its subsequent benefits for both channel parties: the medical practice and their patient panel.

Methodology

To test the relationship of social media to PCM and patient satisfaction, we partnered with a large, urban-based pediatric office in the Northeast United States with a panel of over 10,000 patients. This practice employs three full-time physicians as well as two physician assistants, one nurse practitioner, and several nursing staff. One important consideration in the motivation to improve PCM relationships regards the flexibility of the practice's patients to pursue alternative healthcare providers; that is, for a practice in which the majority of patients have few options or competitors are widely geographically dispersed, it could be argued that there is little incentive to engage in relationship management. In other words, those practices would have a largely captive patient panel. For our study, we interviewed the administrative staff at the medical practice for information on health insurance of their current patient panel. The makeup of the practice's panel was roughly 49% medical assistance (Medicaid) and 51% Highmark Blue Cross/Blue Shield. Both of these insurance options are accepted by all pediatric practices and hospitals in the geographic area (Highmark BC/BS is the insurance of choice for roughly 68% of people in the local tri-county area). With four hospitals, 13 practices, and 32 pediatricians in the metropolitan area, there is, in fact, a high degree of competition among the practices. We further checked to verify that all practices were admitting new patients (they were), so there are also no structural or capacity inhibitors to patients shifting healthcare providers. As a result, the belief of both the practice medical staff and the researchers was that the majority of patients had multiple healthcare options, both from an insurance acceptance and fee perspective (co-pays and percentage cost coverage) and in terms of geographical choice.

The medical practice launched its social media initiative in April 2012, including: a blog and Facebook page regularly updated by the lead physician, a practice website, and a weekly electronic newsletter. Their social media strategy and social media tools were developed by an outreach center at a local university in consultation with the lead physician in the practice.

Six months after the start of the social media program, a survey was conducted to ascertain parent/guardian attitudes toward these initiatives and their impact on PCM and patient satisfaction. The healthcare providers agreed to administer a short survey at the beginning of each office visit. Data were collected in October of 2012, over a two-week period from parents/guardians during only 'well-child appointments.' It was decided that 'sick visits' would be excluded from data collection because the researchers did not want to impose any additional stress on the parents/guardians. All data collection was in strict adherence to university human subject research policies and procedures and the reported data were aggregated and reported in summary form.

Upon entry into the exam room, the provider read a short script to the parents/guardians inviting their participation. The survey and consent form were located in an envelope on the counter in the exam room. If the parent/guardian agreed to participate, the provider would leave the room while the survey was being completed. They were asked to drop their completed survey in a sealed box in the exam room. When finished with the survey, parents/guardians would open the exam room door to indicate they were ready for the provider to return. As an inducement for participation, anyone completing the survey was eligible to participate in a drawing for four tickets to a local water park.

Sample

Two hundred fifty respondents (parents or guardians) answered the survey (male = 13%; female 87%), with a usable sample size of 244. Seventy-one percent of the sample ranged in age from 26 to 45 years of age; with only 11.4% indicating they were under 25 years of age. Sixty-four percent of the respondents were married with 24.1% single, 8.3% divorced, 0.4% widowed, and 3.3% separated. The average number of children in the family was 2.4. In 70% of the families, the father and mother lived in the same household.

The sample was predominately white (222 individuals or 91%). The highest educational level reported was as follows: High school diploma (54 individuals or 22%); some college (60 individuals or 24.5%); College graduate (78 individuals or 31.8%); and Graduate education (47 individuals or 19.2%). For household income, the breakdown was as follows: Less than $30,000 (30.7%); $30,000 to $49,999 (14.5%); $50,000 to $74,999 (22.4%); and over $75,000 (32.4%).

Based on the work of Fox and Jones (2009), we were interested in identifying the percentage of respondents who could be classified as 'e-Patients' – that is, they reported having access to the Internet and going online for health information. Thus, our original usable sample of 244 was pared slightly to the 234 respondents who reported having access to the Internet. Of this sample set, 93% or 218 respondents reported going online for health information for their child. Interestingly, a smaller percentage (88% or 207) of respondents reported going online for health information for themselves. Since this study focuses on pediatric services, we used the 218 respondents (those who reported going online for health information for their child) as our final sample.

Measurement

Internet usage. To assess respondents' Internet usage, several questions were asked: (1) If they had access to the Internet; (2) How they accessed the Internet (computer, smartphone, or tablet); (3) How many minutes/hours per day they spent logged on to the Internet; and (4) How often they used the Internet for various activities, including: checking email, searching for health information for their family, paying bills, and so forth. Items were measured on a 7-point frequency scale that ranged from 1 = Not at all to 7 = Very often.

PCM. There is a wealth of literature on PCM; however, there is little theoretical clarity on the measurement of the construct (Pelzang, 2010; Zandbelt, Smets, Oort, & Haes, 2005). Despite years of research, the PCM construct remains a multi-dimensional construct that continues to be addressed differentially by academic researchers, public policy makers, and consulting organizations. As Zandbelt et al. (2005) note, numerous elements of PCM have been identified, including the physician's communication skills (Stewart et al., 1995); 'perception of patient as person' (Mead & Bower, 2000); quality of the doctor–patient relationship (Stewart et al., 1995), and therapeutic alliance (Mead & Bower, 2000). The Commonwealth Fund, a private foundation that supports research on healthcare issues, acknowledges six key elements of PCM: Education and shared knowledge; Involvement of family and friends, Collaboration and team management; Sensitivity to non-medical and spiritual dimension of care; Respect for patient needs and preferences; and Free flow and accessibility of information (Shaller, 2007, p. 2). Additionally, The Institute For Patient-and Family-Centered Care, a non-profit organization in Maryland that serves as a resource for policy makers, healthcare providers, educators, patients, and families, relies on four core concepts of PCM: Respect and Dignity, Information Sharing, Participation, and Collaboration (Crocker, Webster, & Johnson, 2012).

Recognizing that there is a great deal of overlap between the multiple definitions of PCM, for this study we created a multi-dimensional 21-item measure of PCM that relies on the four core concepts specified by the Institute for Patient-and Family-Centered Care. The individual scale items for these components were adapted from previous research (Beattie, Pinto, Nelson, & Nelson, 2002; Hausman & Mader, 2004; Leisen & Hyman, 2001; Mohr & Spekman, 1994). The items were measured on a 7-point Likert scale that ranged from 1 = Strongly Disagree to 7 = Strongly Agree. Sample items included: 'My pediatrician answers my questions'. 'My pediatrician acknowledges my concerns'. 'My pediatrician asks my input regarding treatment for my child'. We ran a principal components factor analysis on the measure. Results showed that the scale loaded on one dimension (Eigen value = 14.88, Percentage of variance explained = 70.86%), so for the rest of this study, we employed the full 21-item scale to assess PCM. Overall scale reliability, using Cronbach's Alpha, was .98 (see Table A1).

Attitude toward social media. For Attitude toward Social Media, we created a 4-item scale. The items were measured on a 7-point Likert scale that ranged from 1 = Strongly Disagree to 7 = Strongly Agree. Sample items included: 'Social media helps me make better health care decisions for my family'. 'Social media helps me better communicate with my pediatrician.' 'Social media empowers me to make better decisions'. Factor analysis showed a single-dimension scale (Eigen value = 3.27, Percentage of variance explained = 81.8%). The scale reliability (Cronbach alpha) for this construct was .93 (see Table A2).

Patient satisfaction. Finally, for the Patient Satisfaction construct, we employed Welch's (2010) three-dimensional conceptualization: (1) overall satisfaction; (2) likelihood to recommend; and (3) willingness to return. A four-item scale was developed, including sample statements: 'Overall, I am completely satisfied with the care I receive from my pediatrician' and 'I say positive things about my pediatrician to others'. 'I plan to use my pediatrician for all future needs of all my children'. The items were measured on a 7-point Likert scale that ranged from 1 = Strongly Disagree to 7 = Strongly Agree. Factor analysis also demonstrated a single dimension, which accounted for 69.55% of variance explained, with an Eigen value of 2.78. The Cronbach alpha reliability of the satisfaction scale was .84 (see Table A2).

Analysis and results

As part of our analysis, we examined the online activity of the parents/guardians in the medical practice. The respondents with online capabilities indicated that they mainly accessed the Internet via their computer (95%), followed by smartphone (52%), and tablet (25%). For the 10 respondents who indicated 'no access,' they indicated being 'moderately likely' to have access in the future (mean 4.6; scale 1 = not at all likely; 7 = very likely). The primary reasons given for not going online were: hating to use technology (6/10 or 60%) and not being able to afford it (3/10; 30%). Online usage ranged from 0 to 12 hours daily with a mean of 1.65 hours or 99 minutes (S.D. = 110 minutes). A highest frequency of respondents (65 respondents or 26% of the sample) indicated being online approximately 60 minutes per day.

Next, we investigated the primary reasons why respondents accessed the Internet and used social media. Table 1 shows the frequency breakdown for primary uses of social media. The top reasons were to check email ($n = 130$) and pay bills ($n = 72$). However, note that health-related searches for self ($n = 40$) and family members ($n = 52$) were also listed as important reasons for using the Internet.

The correlation table is shown in Table 2. Scale reliabilities are shown on the diagonal. Note that Attitude toward Social Media, PCM, and Patient Satisfaction are all significantly

Table 1. Respondents' online activities.

Online activities	Mean*	Frequency[†]	Percent	Std. Deviation
Checking email	5.2	130	52.9	2.0
Reading someone else's blog	1.9	11	4.5	1.5
Creating or working on your own blog	1.5	10	4.0	1.3
Using Twitter or some other status update	2.0	20	8.2	1.8
Searching for health information for my child/family	3.8	52	21.0	1.8
Searching for health information for myself	3.5	40	16.3	1.7
Looking for and printing coupons	3.0	30	12.2	1.9
Watching online movies or videos	2.6	28	11.4	1.9
Paying bills	3.9	72	29.3	2.2

Note: $n = 239$.
*Scale is measured on a 7-point frequency scale that ranged from 1 = not at all to 7 = very often.
[†]Frequency: # of responses of 6 or 7 on the 7-point scale.

Table 2. Correlation table.

	Mean[†]	S.D.	1	2	3	4
1. Online usage	99 minutes	110 minutes	N.A.			
2. Attitude toward social media	4.46	1.82	.12	(.93)		
3. PCM	6.57	0.75	−.06	.30**	(.98)	
4. Patient Satisfaction	6.50	0.80	.01	.24**	.76**	(.84)

Notes: $n = 239$; Scale reliabilities (Cronbach alpha) on diagonal.
[†]Mean of numbers 2–4 is measured on a 7-point likert scale that ranged from 1 = strongly Disagree to 7 = strongly agree.
**$p < .01$.

correlated with each other ($p < .01$), while online usage shows no significant correlations with any of the other constructs in the study.

We conducted a three-stage regression analysis to establish the path coefficients for each stage in the proposed model. Recall that we proposed a path model to assess general online usage and its impact on both PCM and Patient Satisfaction. Thus, our model tested the impact of online usage and attitude toward social media against the construct of PCM. In the first step, we regressed the exogenous variables (Online Usage and Attitude toward Social Media) against PCM. For the second stage, we then regressed the endogenous, PCM construct against the dependent variable, Patient Satisfaction. Finally, we tested the full model, including both exogenous and endogenous variables. Standardized beta coefficients are used to establish the power and significance of the individual paths in the model. The overall model for our first step, Online Usage and Attitude toward Social Media as predictors of PCM, had a total model r^2 of 0.07, which was significant at the 0.05 level. The full model, including Online Usage, Attitude toward Social Media, and PCM tested as predictors of patient satisfaction, was highly significant ($p < .01$), with a model r^2 of 0.61 (see Table 3).

Table 3 also shows the results of the path model, listing the direct, indirect, spurious, and total effects of antecedent variables on the endogenous variable, PCM, and overall Patient Satisfaction. Note that the construct Attitude toward Social Media had a significant direct effect on PCM and indirect effects on PCM and patient satisfaction. However, neither Attitude toward Social Media nor Online Usage demonstrated a significant direct effect on patient satisfaction, suggesting that the 'correct' chain of causality is through the links,

Table 3. Direct, indirect, and total effects of online usage and attitude toward social media on PCM and patient satisfaction.

	PCM			Patient satisfaction				
Predictors	Direct	Spurious	r	Direct	Indirect	Total	Spurious	r
Online usage	−.09	.03	−.06	.05	.01	.06	−.05	.01
Attitude toward social media	.28**	.02	.30	.06	.02	.08	.16	.24
PCM				.78**		.78	−.02	.76
R^2	.07*					.61**		

*$p < .05$.
**$p < .01$.

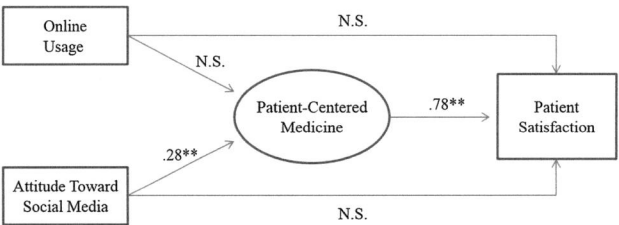

Figure 2. Path model for research constructs.
Notes: $N = 160$, listwise deletion of missing variables, $**p < .01$. Intercorrelations among exogenous and endogenous variables can be found in Table 2.

Attitude toward Social Media → PCM → Overall Patient Satisfaction. Figure 2 demonstrates this mediating role of PCM between Attitude toward Social Media and Patient Satisfaction.

Discussion

Theoretical implications

This article aimed at providing a contribution to healthcare services and related service disciplines pertaining to the impact of social media on customer satisfaction. There were a number of important findings from this study that offer theoretical implications for PCM policy approaches within health service organizations. First, our study confirms the powerful predictive linkage between PCM and patient satisfaction; that is, PCM is a highly significant predictor of subsequent satisfaction, confirming the critical nature of this construct and its usefulness as a means to enhance the quality of the patient–physician dyad. The PCM–Satisfaction link found in our study supports the previous work of Rathert, Wyrwich, and Boren (2013). In their review of the literature on PCM and outcomes, they found in 'almost all studies reviewed, regardless of methodology, a positive relationship between PCM and patient satisfaction' (p. 373). Our study supports the promotion of the PCM meme as an effective approach for improving the critical outcome of patient satisfaction with health service delivery.

Second, there was a significant relationship between patients' Attitude toward Social Media and PCM, ultimately enhancing satisfaction. As Figure 2 illustrates, the original research framework proposed for this paper was partially confirmed through our study, as it demonstrated the significant impact of a patient's positive attitude toward the Internet as a useful means to promote PCM and subsequent satisfaction with healthcare delivery. Put another way, we found that the strong majority of respondents have access to the Internet and frequently use it for a variety of purposes, not the least of which is searching for health information about themselves and family members. The more positive the respondents' attitude was toward the Internet, the more that it can be used as a source for promoting PCM. For example, a recent study by Xie, Wang, Feldman, and Zhou (2013) found that the amount of time spent on the Internet was positively correlated with preferences for obtaining health information and decision making autonomy. Unlike the results of Xie et al. (2013), we did not find evidence of an effect between the simple 'amount' of online usage, as measured in minutes per day on the Internet, and PCM. Being connected, in and of itself, is not the same thing as 'search with a purpose.' Although no direct causal link was found between attitude toward the Internet and patient satisfaction, the mediating role of PCM demonstrates an important endogenous step in this relationship. That is, enhanced patient satisfaction was found to be a function of the fact that the more positive a respondent's attitude is toward the Internet, the greater his/her perception that its use can lead to the characteristics that define the PCM relationship; namely, importance of the patient's perspective, engagement in treatment options, information exchange, and partnership with service provider.

Managerial implications

This study highlights the 'connectedness' of the modern consumer to the web-based and mobile technologies, not only for the more common purposes of checking email and online bill payment, but also as a means for searching for information – for themselves or their family. A number of service industries including legal services (lawyers and legal aid groups), financial services, travel services, athletic training services, and so on are promoting their expertise through social media, publishing expert content on blogs, and distributing content through social networking platforms. These organizations have found that social media is easy to use, increases exposure to consumers in the marketplace, and effectively pushes out information. The common use of Internet applications offers some interesting managerial implications for medical service organizations looking for ways to promote patient education and self-awareness. For example, in the medical practice used in this study, recent initiatives for promoting health information have included physician blogs, Twitter feeds, and Facebook pages, all freely accessible and linked to current medical knowledge and practice information. This case is not unique; hospitals and healthcare organizations are increasingly adopting Internet technologies in order to connect with patients and important community stakeholders, recognizing the efficacy of online systems to link with their services to the broadest possible audience.

> Patient-centered healthcare, social media and the Internet are beginning to come together, with powerful and unpredictable consequences. (Rozenblum & Bates, 2013, p. 183)

As the above reflection makes clear, we are beginning to see a convergence of multiple patterns of social and technology-driven behavior that have powerful implications for the future of healthcare service delivery in this country. In recent years, the benefits of social media have just begun to be felt in medical services, as patients are using it in active searches for themselves and family members, as a means to stay connected with the primary care physicians, and as a means for registering satisfaction (or dissatisfaction) with past interactions with both doctors and healthcare organizations. The medical community must also recognize the risks and challenges that social media presents.

There are professional and legal risks that must be considered. First, social media can distort the border between one's personal and professional life. For example, the Canadian Medical Association recently published issues and rules of engagement for guiding physicians' involvement with social media. The policy states: 'Physicians must retain the appropriate boundaries of the patient-physician relationship when dealing with individual patients. The same standards of professionalism that apply in face-to-face physician-patient interactions also apply in electronic interactions' (Canadian Medical Association, 2011, p. 2). Physicians are warned not to become 'friends and followers' of their patients' sites (e.g. Facebook and Twitter) and thereby allow them access to their personal lives. Medical professionals are cautioned to remember that once something is posted online it is widely accessible. Second, patient confidentiality is paramount. Whatever a physician or other healthcare professional writes online it is their responsibility to abide by all the privacy and security rules of the Health Insurance Portability and Accountability Act

(HIPAA, 1996). Consulting groups such as those from Price Waterhouse Coopers have been formed to assist hospitals, physicians, and other medical providers to develop policies and procedures to govern their social media habits. In sum, social media is embedding itself into the practice of medicine in a very direct way and the implications for this phenomenon are just starting to be felt (PriceWaterhouseCoopers, 2012).

For some years now, the idea of PCM has come to play an increasingly important role in the literature as both a conceptual model and means for enhancing the service dyad between patients and physicians. PCM has been argued to provide many benefits to both patients and providers including better adherence to treatment plans and improved chronic disease control (Stellefson, Dipnarine, & Stopka, 2013; Weiner et al., 2013), treatment effectiveness for pain and emotional problems (Wasson, Johnson, Benjamin, Phillips, & MacKenzie, 2006); better quality of life (Katon et al., 2010), cost effectiveness (Christensen et al., 2013), and decreased utilization of healthcare services (Bertakis & Azari, 2011). This study was undertaken to investigate the impact of social media as a means to enhance the concept of PCM. One of the conundrums facing health service professionals is finding the most effective means to operationalize and leverage PCM as a means for improving the patients' satisfaction with healthcare delivery. With the tremendous accessibility to and use of the Internet, an additional question that must be considered is: How can social media, e-tools, and the Internet add value to and improve the patient (customer) service experience which includes both the PCM experience and ultimately patient satisfaction?

One year post-launch of the social media initiative investigated in this study, a series of follow-up unstructured interviews were conducted with parents/guardians to assess their satisfaction and general attitude about this service enhancement. The nurses/medical assistants solicited comments during the intake procedure of the office visit. This qualitative assessment demonstrated an overwhelmingly positive response from parents/guardians. Sample statements included statements such as:

- 'There is so much to know about when caring for my child, the doctor's blog helps me so much'.
- 'Whenever I have a question, I know where to look for an answer'.
- 'I love your posts on the Facebook page, I share them with friends in my mom's group'.
- 'What you post on social media makes me feel so much better about the things I am trying to do as a mom'.
- 'You give us so many good ideas in your posts. I look for them often. I don't know how you have time for all of this social media work given all of the patients you see everyday'.

While the advantages of a social media program are quickly apparent, the long-term challenge is in the management and maintenance of these initiatives. From a managerial perspective, the veracity of their observation is obvious in analyzing this e-initiative. Specifically, the initial development of the social media program for this medical practice was the result of a service-learning project with a local university. However, there is an issue with staff competency and capability to manage the social media efforts. Currently, the practice's lead physician, who takes personal interest in the project, manages the continual updating of the Facebook page and blog. The practice hired a part-time employee to assist with generating content for the social media tools, but they lack technical expertise for maintenance of the website and identifying and adopting newer and more robust media platforms as they come available. They are still relying on the local university for future maintenance efforts, but purely on an ad hoc basis. As this conversation moves forward, the practice needs to address the resources needed, both human and financial, to make this e-effort successful over the long term.

The study also confirms the efficacy of social media as a means for enhancing the shaping and maintenance of 'network ties' as a critical feature in social capital theory. Specifically, the medical practice in our study embraced Kogut and Zander's (1996) vision of the firm as a social community, emphasizing the speed and efficiency of creating and transferring knowledge through the application of social media methods. For this application to be successful, it required a patient panel that was sufficiently well-versed in the variety of uses and applications of social media as well as one that willingly embraced a refashioning of the traditional doctor–patient dyad, based on limited, problem-based, transactional communications. Redefining the structure of social networks through the use of social media represents an important first step for medical practices that wish to enhance their focus on PCM by highlighting an effective and cost-efficient means to do so.

Finally, it is important to note that social capital derived from online initiatives or social networking sites can be dissipated quickly, unless sufficient time and attention are paid to maintaining these social linkages (Phulari et al., 2010). Put another way, developing online, sharing communities is not a huge challenge but sustaining members' participation can become onerous. The movement away from MySpace and even Facebook for alternative networking platforms illustrates the problem with how users can lose interest in once popular outlets. Healthcare practices must be committed to creating value for participants in their social networks through regular communication and ongoing dissemination of fresh and relevant information, updating blogs, responding to queries, and supplying requested. These activities require continuous support and sufficient resources in terms of people, time, and money. Once these sites are established to promote continued PCM, they must be maintained in a manner that continues to provide value to all participants.

Limitations and future research

In addition to these findings, some limitations of the current study should be considered. The sample, which consisted of data collected from only one urban medical practice in a northeastern city, may be a potential limitation. Sampling protocol would be enhanced by identifying and sampling from multiple sites in different locations, perhaps also from practices specializing in different branches of medicine and with a broader array of patients. For example, as a pediatric practice, we found that the majority of parents (71%) were in the 26–45-year-old demographic; a group that has grown up with and become familiar with the Internet and are proficient in search procedures and other utilities. It would be useful and instructive to identify other age demographics to ascertain their general use of the Internet and social media, the purposes for which they most often use their time online, and its impact on the Attitude toward the Internet–PCM–Satisfaction linkage.

Secondly, we examined the attitude toward social media as a general construct, using broad-based questions such as 'Social media helps me better communicate with my pediatrician.' While these questions offer a general perspective on the respondents' attitude toward the use of social media as a communication device, they do not parse out the different platforms that can be used; for example, Facebook, Twitter, LinkedIn, and so forth. For a medical practice that is interested in which of these tools can be most useful for communicating with their patient panel, it would be instructive to sharpen this analysis in order to identify the strengths and weaknesses of each of these applications for promoting PCM. Future research could break down the construct of 'Attitude toward Social Media' into more specific categories to determine the effectiveness of alternative means to enhance PCM and satisfaction.

This research was limited geographically and focused on one large medical practice located in the USA. The US healthcare model differs from that of many other countries, with characteristics of incentives and market-based competition between medical practices. This competitive model suggests that it is to a medical practice's advantage to facilitate a PCM approach in an effort to maintain their patient panel and attract new patients. However, it may be the case that some PCM behaviors and motivations are context (country) specific; that is, it is likely that joint patient–physician motivation to adopt PCM in the US market is a function of its competitive nature. In other countries, where healthcare choice is more limited and the physician to whom you are assigned is the physician you must see, it may be the case that expectations of a PCM attitude by both the patient and the medical practice are much diminished. Even in the USA, there are locales where a patient's choice of healthcare provider is limited by their insurance plans to in-network options or the patient must be willing to bear higher costs to receive these out-of-network medical services. Future research on the use of social media to enhance PCM and ultimately patient satisfaction should address these attitudes in a cross-regional or multi-national setting. Internationally, is social media a useful tool for enhancing PCM and, indeed, to what degree is PCM an explicit or even implicit goal of these healthcare systems? Additionally, testing physicians' attitudes toward social media and the degree to which PCM is a compelling goal for them would offer an interesting complementary perspective to this study.

In the USA, where this study was conducted, there is a strong motivation to pursue the goals of PCM. The development of the Affordable Care Act ('Obamacare') includes a number of salient markers for quality, including patient satisfaction.

> The Patient Protection and Affordable Care Act of 2010 (Obamacare) creates several new Medicare programs intended to improve health care quality, using 'pay-for-performance' payment strategies to put financial pressure on medical providers. In such programs, reimbursement reflects provider performance on metrics based on adherence to certain care processes, scores on patient satisfaction surveys, or patient outcomes. (Nix, 2013, p. 1)

Thus, satisfaction and drivers of satisfaction (in our study, the PCM model) are a critical metric for patient reimbursement and therefore a major motivator to the medical provider.

Voluminous research over the past decades has identified numerous factors that can contribute to patient satisfaction; further, the causal link between patient satisfaction and PCM has frequently been elaborated and explored. Indeed, since the Institute of Medicine's *Crossing the Chasm* (2001) arguments spoke so strongly in favor of the pursuit of PCM, healthcare organizations have accepted this call and sought to find the means to identify the critical factors that create a PCM philosophy while working to enhance their promotion in the medical workplace. This study examined one means by which health service providers can exploit the PCM–patient satisfaction linkage through employing social media as an effective tool for promoting PCM. Future research should continue to highlight the increasing role of social media as well as the types of information that can best be conveyed by these technologies as healthcare grapples with the best methods by which to pursue the worthy goal of PCM.

Disclosure statement
No potential conflict of interest was reported by the authors.

References

Adler, P. S., & Kwon, S.-W. (2002). Social capital: Prospects for a new concept. *Academy of Management Review, 27*, 17–40.

Anheier, H. K., Gerhards, J., & Romo, F. P. (1995). Forms of capital and social structure in cultural fields: Examining Bourdieu's social topography. *American Journal of Sociology, 100*, 859–903.

Bacigalupe, G. (2011, March). Is there a role for social technologies in collaborative healthcare? *Families, Systems, & Health, 1*, 1–14.

Balint, E. (1969). The possibilities of patient-centered medicine. *Journal of the Royal College of General Practicioners, 17*(82), 269–276.

Beattie, P. F., Pinto, M., Nelson, M. K., & Nelson, R. (2002). Patient satisfaction with outpatient physical therapy: Instruments validation. *Physical Therapy, 82*(6), 557–564.

Bertakis, K., & Azari, R. (2011). Patient-centered care is associated with decreased health care utilization. *The Journal of the American Board of Family Medicine, 24*(3), 229–239.

Bourdieu, P., & Wacquant, L. (1992). *An invitation to reflexive sociology*. Chicago, IL: University of Chicago Press.

Byrne, P., & Long, B. (1976). *Doctors talking to patients*. London: HMSO.

Canadian Medical Association. (2011). *Social media and Canadian physicians – issues and rules of engagement*. Ottawa, ON: CMA.

Christensen, E. W., Dorrance, K., Ramchandani, S., Lynch, S., Whitmore, C. C., Borsky, A. E., … Bicket, T. A. (2013). Impact of a patient-centered medical home on access, quality, and cost. *Military Medicine, 178*(2), 135–141.

Crocker, E., Webster, P. D., & Johnson, B. H. (2012). *Developing family-centered vision, mission, and philosophy of care statements*. Bethesda, MD: Institute for Patient-and Family-Centered Care.

Donath, J., & Boyd, D. (2004). Public displays of connection. *BT Technology Journal, 22*(4), 71–82.

Ellison, N. Steinfield, C., & Lampe, C. (2007). The benefits of Facebook 'friends:' social capital and college students' use of online social sites. *Journal of Computer Mediated Communication, 12*(4), 1143–1168.

Epstein, K. R., Laine, C., Farber, N. J., Nelson, E. C., & Davidoff, F. (1996). Patients' perceptions of office medical practice: Judging through the patients' eyes. *American Journal of Medical Quality, 11*(2), 73–80.

Fox, S., & Jones, S. (2009). *The social life of health information*. Washington, DC: Pew Research Center.

Graham, D. L. (2011). Social media and oncology: Opportunity with risk. *American Society of Clinical Oncology Educational Book, 31*, 422–424.

Granitz, N., & Koernig, S. (2011). Web 2.0 and marketing education: Explanations and experiential applications. *Journal of Marketing Education, 33*(11), 57–72.

Hausman, A. V., & Mader, D. (2004). Measuring social aspects in the physician/patient relationship. *Health Marketing Quarterly, 21*(3), 3–26.

Health Insurance Portability and Accountability Act (HIPAA). (1996). Pub. L. No. 104–191, 110 Stat 1936.

Hojat, M., Louis, D. Z., Maio, V., & Gonnella, J. S. (2013). Empathy and health care quality. *American Journal of Medical Quality, 28*(1), 6–7.

Institute of Medicine (2001). *Crossing the quality chasm: A new health system for the 21st century*. Washington, DC: National Academy Press.

Katon, W. J, Lin, E. H. B., Von Korff, M., Ciechanowski, P., Ludman, E. J., Young, B., … McCulloch, D. (2010). Collaborative care for patients with depression and chronic illnesses. *New England Journal of Medicine, 363*(27), 2611–2620.

Kim, S. S., Kaplowitz, S., & Johnson, M. V. (2004). The effects of physician empathy on patient satisfaction and compliance. *Evaluation and the Health Professions, 27*(3), 237–251.

Kogut, B., & Zander, U. (1996). What firms do? Coordination, identity, and learning. *Organization Science, 7*, 502–518.

Kraschnewski, J. L., Chuang, C. H., Poole, E. S., Peyton, T., Blubaugh, I., Pauli, J., … Reddy, M. (2014). Paging "Dr. Google": Does technology fill the gap created by the prenatal care visit structure? Qualitative focus group study with pregnant women. *Journal of Medical Internet Research, 16*(6), e147. doi:10.2196/jmir.3385

Lane, S. H. (2010). Should social media be used to communicate with patients? *The American Journal of Maternal/Child Nursing*, *35*(1), 6–7.

Leisen, B., & Hyman, M. R. (2001). An improved scale for assessing patients' trust in their physician. *Health Marketing Quarterly*, *9*(1), 23–42.

Lin, N., Cook, K., & Burt, R. S. (2008). *Social capital: Theory and research* (4th ed.). New Brunswick, NJ: Transaction Publishers.

Little, P., Everitt, H., Williamson, I., Warner, G., Moore, M., Gould, C., … Payne, S. (2001). Observational study of effect of patient centeredness and positive approach on outcomes of general practice consultations. *British Medical Journal*, *323*, 908–911.

Mathwick, C., Wiertz, C., & de Ruyter, K. (2008). Social capital production in a virtual P3 community. *Journal of Consumer Research*, *34*, 832–849.

McDaniel, B. T., Coyne, S. M., & Holmes, E. K. (2011). New mothers and media use: Associations between blogging, social networking, and maternal well-being. *Maternal and Child Health Journal*, *16*(7), 1509–1517.

Mead, N., & Bower, P. (2000). Measuring patient-centeredness: A comparison of three observations-based instruments. *Patient Education and Counseling*, *39*(1), 71–80.

Mohr, J., & Spekman, R. (1994). Characteristics of partnership success: Partnership attributes, communication behavior, and conflict resolution techniques. *Strategic Management Journal*, *15*(2), 135–152.

Morrow, V. (1999). Conceptualizing social capital in relation to the well-being of children and young people: A critical review. *Sociological Review*, *47*, 744–765.

Nielsen. (2011). *The state of the media: The social media report Q3 2011*. New York, NY: Neilson.

Nix, K. (2013). *What Obamacare's pay-for-performance programs mean for health care quality (Backgrounder No.2856)*. Washington, DC: The Heritage Foundation.

Norrish, A., Biller-Andorno, N., Ryan, P., & Lee, T. H. (2013, November 20). Social capital is as important as financial capital in health care. *Harvard Business Review*. Retrieved January 8, 2015, from https://hbr.org/2013/11/social-capital-is-as-important-as-financial-capital-in-health-care/

Pelzang, R. (2010). Time to learn: Understanding patient-centred care. *British Journal of Nursing*, *19*(14), 912–917.

Phulari, S. S., Khamitkar, S. D., Deshmukh, N. K., Bhalchandra, P. U., Lokhande, S. N., & Shinde, A. R. (2010). Understanding formulation of social capital in online social network sites (SNS). *International Journal of Computer Science Issues*, *7*(1), 92–96.

PricewaterhouseCoopers, Health Research Institute. (2012). *Social media 'likes' healthcare: From marketing to social business*. New York, NY: PricewaterhouseCoopers.

Rathert, C., Wyrwich, M. D., & Boren, S. A. (2013). Patient-centered care and outcomes: A systematic review of literature. *Medical Care Research and Review*, *70*(4), 351–379.

Rozenblum, R., & Bates, D. W. (2013). Patient-centered healthcare, social media and the Internet: The perfect storm? [Editorial]. *BMJ Quality & Safety*, *22*(3), 183–186.

Safko, L. (2010). *The social media bible: Tactics, tools, and strategies for business success*. Hoboken, NJ: John Wiley.

Saultz, J. W., & Lochner, J. (2005). Interpersonal continuity of care and care outcomes: A critical review. *Annals of Family Medicine*, *3*(2), 159–166.

Shaller, D. (2007, October). *Patient-centered care: What does it take?* Washington, DC: The Commonwealth Fund.

Solomon, M. R. (2013). *Consumer behavior: Buying, having, and being* (10th ed.). Boston: Prentice Hall.

Steinfield, C., Ellison, N. B., & Lampe, C. (2008). Social capital, self-esteem, and use of social network sites: A longitudinal analysis. *Journal of Applied Developmental Psychology*, *29*, 434–445.

Stellefson, M., Dipnarine, K., & Stopka, C. (2013, February). The chronic care model and diabetes management in US primary care settings: A systematic review. *Preventing Chronic Disease*, *10*, 1–21.

Stewart, M., Brown, J. B., Donner, A., McWhinney, I. R., Oates, J., Weston, W. W., & Jordan, J. (2000). The impact of patient-centered care on outcomes. *Journal of Family Practice*, *4*(9), 796–804.

Stewart, M., Brown, J., Weston, W., McWhinney, I., McWilliam, C., & Freeman, T. (1995). *Patient-centered medicine: Transforming the clinical method*. London: Sage.

Szreter. S., & Woolcock, M. (2004). Health by association? Social capital, social theory, and the political economy of public health. *International Journal of Epidemiology*, *33*, 650–667.

Thielst, C. (2011). Social media: Ubiquitous community and patient engagement. *Public Health and Safety, Health Facilities and Administration, Business and Economics*, *28*(2), 3–14.

Wasson, J., Johnson, D., Benjamin, R., Phillips, J., & MacKenzie, T. (2006). Patients report positive impacts of collaborative care. *Journal of Ambulatory Care Management*, *29*(3), 199–206.

Weiner, S. J., Schwartz, A., Sharma, G., Binns-Calvey, A., Ashley, N., Kelly, B., … Harris, I. (2013). Patient-centered decision making and health care outcomes: An observational study. *Annals of Internal Medicine*, *158*(8), 573–579.

Welch, S. (2010). Twenty years of patient satisfaction research applied to the emergency department: A qualitative review. *American Journal of Medical Quality*, *25*(1), 64–72.

Xie, B., Wang, M., Feldman, R., & Zhou, L. (2013). Internet use frequency and patient-centered care: Measuring patient preferences for participation using the health information wants questionnaire. *Journal of Medical Internet Research*, *15*(7), 132–157.

Zandbelt, L. C., Smets, E. M. A., Oort, F. J., & Haes, H. (2005). Coding patient-centered behavior in the medical encounter. *Social Science and Medicine*, *61*(3), 661–671.

Zickuhr, K. (2013, September 13). *Who's not online and why*. Washington, DC: Pew Research Center.

Appendix

Table A1. Factor analysis of PCM scale.

Scale	Item	Components*
PCM	PCM (Cronbach alpha = 0.98)	1
	My pediatrician …	
	Answers my questions	0.778
	Gives me complete information that helps me care for my child	0.682
	Gives me clear instructions about when to call the office or seek further care for my child	0.770
	Listens carefully to me	0.789
	Is courteous	0.868
	Treats me with respect	0.894
	Is sensitive to my feelings	0.876
	Acknowledges my concerns	0.903
	Is non-judgmental toward my family and me	0.814
	Advises me	0.850
	Asks my input regarding treatment options for my child	0.743
	Works with me to set goals and discuss treatment options for my child	0.835
	Accepts me for who I am	0.875
	Treats me as an equal	0.863
	Is open to new ideas	0.839
	Is concerned about the comfort of my child	0.854
	Expresses concern for the well-being of my child	0.887
	Is empathetic	0.866
	Is reassuring and comforting	0.894
	Is hopeful	0.915
	Expresses concern for my well-being	0.840
	Eigen value = 14.88	
	Percentage variance explained = 70.86%	

*Only one component was extracted. Only the component matrix is presented because the solution cannot be rotated.
Note: Extraction method: principal component analysis.

Table A2. Factor analysis of attitude toward social media and patient satisfaction scales.

Scale	Items	Components
Attitude toward social media*	Attitude toward social media (Cronbach alpha = 0.93)	1
	Social media helps me make better healthcare decisions for my family	0.942
	Social media empowers me to make better decisions	0.952
	I am likely to share information I learned through social media with my family and friends	0.765
	Social media helps me better communicate with my pediatrician	0.945
	Eigen value = 3.27	
	Percentage variance explained = 81.8%)	
Patient satisfaction*	Patient satisfaction (Cronbach alpha = 0.84	1
	Overall, I am completely satisfied with the care I received from my pediatrician	0.845
	I say positive things about my pediatrician to others	0.865
	I encourage my friends and family members to bring their children to my pediatrician	0.823
	I plan to use my pediatrician for all the needs of my children	0.799
	Eigen value = 2.78	
	Percentage variance explained = 69.55%	

*Only one component was extracted. Only the component matrix is presented because the solution cannot be rotated.
Note: Extraction method: principal component analysis.

Extending customer relationship management into a social context

Sarah Diffley[a] and Patrick McCole[b]

[a]Letterkenny Institute of Technology, Letterkenny, Co. Donegal, Ireland; [b]Queen's University Management School, Queen's University Belfast, Belfast, Northern Ireland

Informed by the resource-based view, this study draws on customer relationship management (CRM) and value co-creation literature to develop a framework examining the impact of social networking sites on processes to manage customer relationships. Facilitating the depth and networked interactions necessary to truly engage customers, social networking sites act as a means of enhancing customer relationships through the co-creation of value, moving CRM into a social context. Tested and validated on a data set of hotels, the main contribution of the study to service research lies in the extension of CRM processes, termed relational information processes, to include value co-creation processes due to the social capabilities afforded by social networking sites. Information technology competency and social media orientation act as critical antecedents to these processes, which have a positive impact on both financial and non-financial aspects of firm performance. The theoretical and managerial implications of these findings are discussed accordingly.

Introduction

Customer relationship management (CRM) literature has long emphasised the necessity to create co-value as a means of developing and maintaining customer relationships, achieved through the process of interacting with customers and managing the information garnered from those interactions (Boulding, Staelin, Ehret, & Johnston, 2005; Payne & Frow, 2005). However, while CRM focuses on developing and maintaining profitable, mutually beneficial customer relationships through the co-creation of value (Boulding et al., 2005; Ernst, Hoyer, Krafft, & Krieger, 2011; Payne & Frow, 2005), the results of CRM initiatives have been mixed, with many failures reported (Jayachandran, Sharma, Kaufman, & Raman, 2005; Zablah, Bellenger, & Johnston, 2004). Marketers are increasingly paying attention to social media technologies as a means of overcoming the obstacles encountered in implementing effective CRM initiatives, thus extending CRM into a social CRM context (Greenberg, 2010; Gummesson & Mele, 2010; Trainor, 2012; Trainor, Andzulis, Rapp, & Agnihotri, 2014).

Informed by the resource-based view (RBV), this study proposes a model that applies and extends the relational information processes developed by Jayachandran et al. (2005) to investigate the impact of social networking sites on organisational routines necessary for

CRM. The social capabilities afforded by social networking technologies, when combined with the interaction and information management capabilities inherent in relational information processes, are proposed to facilitate the co-creation of value with customers. That is, social networking sites are anticipated to offer the depth and networked interactions necessary to engage customers in the co-creation of value, reflecting the true nature of customer relationships and hence CRM.

The study was conducted in the Irish hotel industry. Given that customer experiences are central to the tourism and hospitality industry, the relationship that exists between firms and customers has become increasingly interactive, providing opportunities for value co-creation. Accessing customer information necessary to co-create value can occur in several ways in the hotel industry, including guest comment cards, questionnaires, trials, face-to-face discussions and gauging customer reactions (Shaw, Bailey, & Williams, 2011). Social networking sites are becoming increasingly important as a means of accessing this valuable customer information in the hotel industry (Shaw et al., 2011; Sigala, 2009). The output of value co-creation activities can extend to numerous aspects of hotel operations, including refurbishment activities, in-room technologies, room design, customer dining experiences and website booking engines.

Putting the 'social' in CRM

A review of literature indicates three distinct but related reasons explaining the mixed successes of CRM initiatives. First, Gummesson (2006) attributes the one-to-one focus of CRM initiatives as a reason for the difficulties and failures encountered by companies. The one-to-one or dyadic view fails to account for the widened view of relationships that typically involves more than two customers or companies, leading to a networked or many-to-many perspective (Gummesson & Mele, 2010). Second, the lack of interactivity delivered by traditional CRM technologies, which typically facilitate one-way monologues, have inhibited the ability of companies to engage with customers in these networked relationships (Trainor, 2012). Third, the belief that investment in CRM technology alone is enough to act as a means of improving performance is impractical (Fan & Ku, 2010). Improved performance is achieved through a process of combining technological and additional resources, which, due to their complementary nature, result in the development of distinctive capabilities (Trainor et al., 2014). In effect, CRM technologies result in enhanced customer relationships when applied to customer-centric business processes (Jayachandran et al., 2005; Rapp, Trainor, & Agnihotri, 2010), a perspective that is aligned with that of the RBV (Coltman, 2007a; Rapp et al., 2010).

Social CRM extends the concept of CRM to include 'the integration of traditional customer-facing activities including processes, systems, and technologies with emergent social media applications to engage customers in collaborative conversations and enhance customer relationships' (Trainor, 2012, p. 319). The high-level company–customer interactions afforded by social technologies provide a means of more effectively engaging and co-creating value with customers, resulting in a more complete picture of customers and their networks being obtained. This is achieved as social CRM technologies capture the networked, many-to-many reality of customer relationships not previously possible with traditional CRM technologies. Hence, the collaborative activities afforded by social CRM technologies move customer relationships towards a process of engaging rather than managing people (Greenberg, 2010).

Theoretical background and hypotheses

The RBV proposes that the true source of competitive advantage for a firm lies in its ability to acquire and control valuable resources that are rare and inimitable. These resources extend to the management skills, processes and routines, and information and knowledge that exist within a firm, facilitating greater levels of efficiency and effectiveness (Barney, 1991). Resources may then be combined in a manner that facilitates the development of capabilities, defined by Rapp et al. (2010, p. 1229) as 'a firm's ability to assemble, integrate and deploy valuable resources in combination to achieve superior performance'. Hence, capabilities represent the purposeful, knowledge-based combination of resources in a certain manner in order to develop complex and inimitable capabilities that provide a means of gaining competitive advantage (Zablah et al., 2004). The RBV has garnered significant attention in marketing strategy, information technology (IT) and information systems literature in order to determine the specific mix of resources, and hence capabilities that are necessary in order to achieve superior levels of firm performance; given that a focus on technological resources alone has been found insufficient as a means of gaining a competitive advantage (Coltman, 2007a; Rapp et al., 2010). Therefore, it is evident that CRM literature is nested within the RBV (Coltman, 2007a; Kermati, Mehrabi, & Mojir, 2010; Rapp et al., 2010) and that the success of CRM initiatives extends beyond the deployment of technological resources alone to the combination of technological, human and business-related resources. It is the combination of these resources that facilitate the development of capabilities in managing customer relationships (Coltman, 2007a). Dynamic capabilities literature extends the RBV to address the dynamic nature of the business environment. Dynamic capabilities represent 'the firm's ability to integrate, build and reconfigure internal and external competences to address rapidly changing environments' (Teece, Pisano, & Shuen, 1997, p. 516). Thus, the routines or processes that reconfigure resources act as the key source of competitive advantage (Teece, Pisano, & Shuen, 1997). The importance of reconfiguring knowledge-based resources has been stressed in the literature (Eisenhardt & Martin, 2000). Knowledge facilitates learning and learning is inherently social in nature (Teece et al., 1997). Literature draws attention to the role of digital resources in facilitating dynamic capabilities. Bharadwaj, El Sawy, Pavolu, and Venktraman (2013, p. 472) cite the example of Susarla, Oh, and Tan (2012) to exemplify how digital technologies are 'transforming the structure of social relationships in both the consumer and enterprise space with social media and social networking'.

Both CRM and value co-creation literature view the co-creation of value as a means of gaining a competitive advantage and it is only through interactive relationships that this co-creation of value is achieved (Boulding et al., 2005; Vargo & Lusch, 2004). Interactions allow parties to communicate their needs and wants, which in turn, creates a foundation for the exchange of valuable resources. These resources can be intangible and/or tangible in nature and, through meeting the needs and wants of the parties involved in the exchange, serve to create and maintain stronger relationships. As such, interactions rather than goods act as the core of relationships, shifting the focus of value from one of value exchange to value creation (Gummesson & Mele, 2010). Interactions facilitate communication and it is through this communication that exchange takes place, including the exchange of information between a firm and its customers. This information is essential in developing customer relationships and it is 'imperative that organizations use the information to shape appropriate responses to customer needs. In effect, information plays a key role in building and maintaining customer relationships' (Jayachandran et al., 2005, p. 178). This suggests that intangible resources are essential in developing customer relationships and co-creating value with them. A firm can easily invest in the same CRM technologies as competitors, which alone does not result in the same or higher levels of performance being achieved. It is when inimitable, intangible CRM resources are combined with tangible CRM resources that a sustainable competitive advantage is achieved (Kim, Kim, & Park, 2010).

This view of value co-creation is in line with that stressed by the service-dominant (S-D) logic, first introduced by Vargo and Lusch (2004). The S-D logic views that it is intangible 'operant resources' that act as a means of gaining a competitive advantage. Knowledge and skills represent these operant resources. This emphasises that goods (operand resources) are not sufficient in gaining a competitive advantage. They must be acted upon by operant resources in order to product an effect (Vargo & Lusch, 2006). Hence, firms are concerned with the exchange of service – the 'application of specialised competencies (operant resources – knowledge and skills) through deeds, processes, and performances for the benefit of another entity or the entity itself' (Vargo & Lusch, 2006, p. 43), thereby shifting the focus of value to those processes that integrate and transform resources. Accordingly, value is co-created by producers and consumers through the integration of resources and application of competences (Lusch & Vargo, 2006). A firm cannot deliver value but only offer value propositions, which if accepted by the customer leads to the co-creation of value (Vargo & Lusch, 2004). The enactment of value propositions is known as value-in-use and acts as a driver in the value co-creation process. Value-in-use has been extended to the more S-D logic friendly term of value-in-context to acknowledge that value is determined by the customer based on contextual factors (Vargo, Maglio, & Akaka, 2008). The importance of networks is also implied within the S-D logic. Networks act as links between buyers and sellers (Lusch & Vargo, 2006). In order for value to be co-created interactive, reciprocal relationships are inferred (Ballantyne & Varey, 2006). As the process of value co-creation requires the integration, transformation and application of resources from various parties, these relationships occur in networks (Vargo, 2008), thus emphasising many-to-many marketing communications (Cova & Salle, 2008).

The significant amount of research in the CRM domain has resulted in the emergence of five divergent perspectives of CRM: CRM as a process, strategy, philosophy, capability and technology. While each perspective provides valuable insights into the CRM concept, it is the process perspective that is advocated as the most appropriate CRM lens as it acknowledges the changing and evolving nature of buyer–seller relationships (Zablah et al., 2004). The process perspective of CRM has been adopted in numerous research studies (e.g. Ernst et al., 2011; Jayachandran et al., 2005; Payne & Frow, 2005; Reinartz, Krafft, & Hoyer, 2004). A process perspective of social CRM is adopted in the current study, with the proposed conceptual model shown in Figure 1.

Process perspective of social CRM

The CRM process perspective views resources as inputs that are transformed in a manner that allows desired outputs to be achieved. Therefore, resources play a critical role in the CRM process (Zablah et al., 2004). As emphasised by the RBV, it is the unique combination of resources that allow distinctive capabilities to be developed, thereby providing a competitive advantage (Coltman, 2007a; Rapp et al., 2010). At a process level, intangible complex resources that are difficult to identify and describe act as a means of developing the capabilities necessary to build and maintain profitable, mutually beneficial customer relationships (Raman, Wittman, & Rauseo, 2006). As resources are enablers of CRM processes, 'CRM capabilities can be best described at the process level' (Kermati et al., 2010, p. 1177).

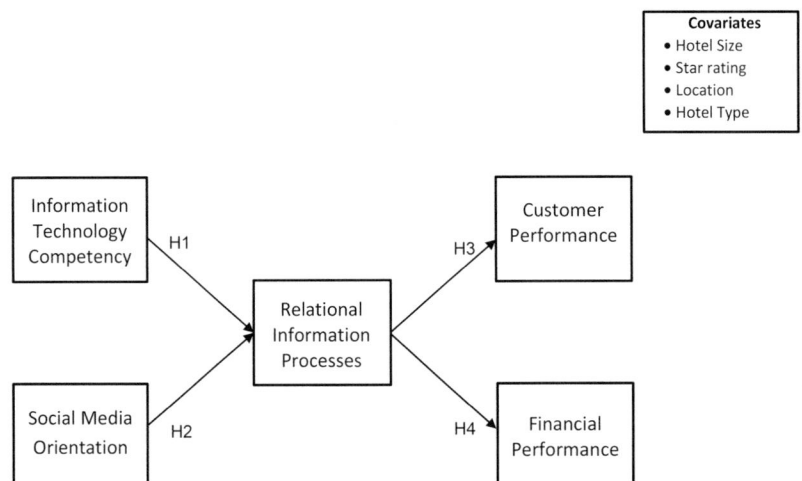

Figure 1. Conceptual model.

The centrality of customer interaction and information management in the CRM process is stressed in CRM literature (Boulding et al., 2005; Fan & Ku, 2010; Garrido-Moreno & Padilla-Meléndez, 2011; Jayachandran et al., 2005; Mithas, Krishnan, & Fornell, 2005; Payne & Frow, 2005). Hence, CRM processes should be designed to facilitate close customer–company interactions (Raman et al., 2006). The collection of information from these interactions may then be processed such that customer knowledge can be generated and applied to respond to customer needs, regardless of context (Mithas et al., 2005). Consequently, the management of information is also essential as 'to collect information about customers in the context of a relationship, and offer those customers a superior value proposition based on this knowledge, will be a key advantage, hard to imitate' (Garrido-Moreno & Padilla-Meléndez, 2011, p. 438).

Similarly, value co-creation literature stresses the importance of customer interactions and information. Interaction, as the locus of value co-creation, allows dialogue to be established and the transfer of operant resources (skills and knowledge) to take place. This forms the basis for value co-creation. The quality of interactions between a firm and its customers is also important as it has a direct impact on motivation and ability to co-create value, implying that interactions must be managed effectively (Fyrberg & Jüriado, 2009). Value co-creation literature also draws attention to the exchange and application of knowledge in the co-creation process. Ballantyne and Varey (2006) note that knowledge renewal: the process of generating, sharing and applying knowledge is a more accurate description of the role played by knowledge in gaining a competitive advantage. To derive value from knowledge, it must be acted upon. As knowledge and knowledge renewal are central to the co-creation of value, knowledge management practices in a firm should be designed around value co-creation, and hence knowledge renewal processes (Payne, Storbacka, & Frow, 2008). As emphasised by Echeverri and Skålén (2011, p. 354), 'it is only when the knowledge and skills, or the operant resources, are active or activated that value co-creation takes place'.

Based on these findings, this study determines that information and interaction management represent the capabilities necessary to execute the CRM process in order to build and maintain customer relationships that allow the co-creation of value to take place. Following the need to integrate CRM technologies with CRM processes, it can be deduced that social CRM technologies must be integrated with existing CRM technologies and processes as a means of improving customer relationships, and hence firm performance (Boulding et al., 2005; Jayachandran et al., 2005; Payne & Frow, 2005; Rapp et al., 2010; Zablah et al., 2004). Through the provision of the social capabilities necessary to truly interact and engage with customers in a networked context, social CRM resources act as a means of gaining a competitive advantage through the co-creation of value with customers.

Jayachandran et al. (2005) describe the process of interacting with customers and managing customer information to establish long-term relationships as relational information processes. Originally consisting of five dimensions of information reciprocity, information capture, information integration, information access and information use, these dimensions represent the organisational routines essential for CRM. This research adds a sixth dimension, value co-creation, to the original conceptualisation. This is warranted as the social capabilities afforded by social networking sites, when integrated with these dimensions, facilitate the depth interactions necessary to co-create value with customers, accounting for social CRM processes.

Dimensions of relational information processes

Information reciprocity

Information reciprocity refers to the processes that facilitate mutual, high-level information exchanges to take place between a firm and its customers. Interaction and dialogue are aspects of information reciprocity (Ballantyne & Varey, 2006; Prahalad & Ramaswamy, 2004; Ramaswamy & Gouillart, 2010). Interaction facilitates dialogue which in turn facilitates the shaping of value propositions and the exchange of important information (operant resources) which may then be applied in the co-creation of value (Gummesson & Mele, 2010). The depth and networked interactions facilitated by social networking sites provide the additional customer insights that traditional CRM technologies have been lacking (Greenberg, 2010).

Information capture

Comprehensive and current customer information must be obtained from customer interactions if customer relationships are to be developed (Jayachandran et al., 2005). This information must also be kept up-to-date to offset any change in the definition of value from any stakeholder in the value network (Lusch & Webster, 2011). In light of the vast amounts of valuable customer information available via social networking sites, capturing information from these sites is essential.

Information integration

It is necessary to integrate customer information from the various sources that interact with customers in order to develop a comprehensive account of customer relationships and to ensure that customer information is not lost. Not only must customer information be collected across customer touch points but it must also be collated if a true understanding of customers is to be achieved (Payne & Frow, 2005). This extends to the integration of information collected from social networking sites with information collected via traditional CRM technologies (Trainor, 2012).

Information access

Those employees who actively develop value propositions and co-create value with customers must have access to current and full customer information in order to do so effectively (Lusch & Webster, 2011). Also, an organisation must capture and apply the information garnered from co-creative interactions in order to improve future co-creative experiences (Karpen, Bove, & Lukas, 2012). In order to achieve this, it can be inferred that access to the appropriate information by employees who undertake co-creative activities is necessary, including those who engage with customers via social networking sites.

Information use

The information acquired from customer interactions must be applied if it is to aid in understanding customer needs, wants and behaviours (knowledge-enhancing use) and lead to the development of market offerings that meet these needs and wants (action-oriented use) (Jayachandran et al., 2005). In order to co-create value, information from dialogical interactions must first be applied in order to develop and maintain customer relationships. The use of this information results in customer engagement, the creation of meaningful, deep and long-term connections between a firm and its customers (van Doorn et al., 2010). It is within states of customer engagement that the co-creation of value occurs (Brodie, Hollebeck, Juric, & Ilic, 2011; van Doorn et al., 2010). Consequently, the information acquired from customer interactions via social CRM and traditional CRM technologies must be applied if it is to provide knowledge-enhancing and action-oriented use, which both serve to build and maintain profitable, mutually beneficial customer relationships within which the co-creation of value takes place.

Value co-creation

Value co-creation represents 'a firm's efforts to interact with customers to co-construct their consumption experiences' (O'Cass & Ngo, 2011, p. 652). The process of value co-creation is such that the customer is central to the co-creation experience (Prahalad & Ramaswamy, 2004). To allow customers to co-create their own experiences, firms must provide the necessary assistance and support (Karpen et al., 2012). In effect, firms should strive to be an efficient and effective service support system (Lusch & Webster, 2011) as it is the responsibility of the firm to create and manage value co-creation opportunities. The business strategy must be one of understanding the processes by which customers create value and determining which of these processes they will support. The firm then aims to support customer processes so that better value may be co-created (Payne et al., 2008). As it is within states of customer engagement that the co-creation of value takes place (Brodie et al., 2011; van Doorn et al., 2010) information must first be applied to build and maintain customer relationships. Upon the development of these relationships, customer information may then be applied as a means of co-creating value with customers. As full, comprehensive and consistent information is necessary to develop customer relationships (Jayachandran et al., 2005), it is the integration of information from social CRM and traditional CRM technologies that provides the information necessary to engage and hence co-create value with customers.

Antecedents to relational information processes

Drawing on the RBV, it is the purposeful combination of technological, human and business resources that facilitates the development of capabilities necessary to manage customer relationships as a means of gaining a competitive advantage. The sub-processes comprising relational information processes represent business resources that combine to facilitate the development of interaction and information management capabilities. When the social capabilities facilitated by social networking sites are combined with these capabilities, the sub-processes comprising relational information processes can be extended to include value co-creation processes. It is proposed that IT competency and social media orientation act as critical antecedents in developing relational information processes.

IT competency

IT competency reflects the knowledge and use of IT as a means of managing information within a firm. The three components of an IT competency relate to the IT knowledge of staff, IT skills of staff and the quality of IT infrastructure within a firm (Tippins & Sohi, 2003). Thus, IT competency represents the combination of human and technological CRM resources (Coltman, 2007a; Rapp et al., 2010). CRM literature communicates the importance of the appropriate IT infrastructure and the requisite skills and knowledge among staff to effectively use physical IT resources to interact with customers and manage customer information (Coltman, 2007a; Rapp et al., 2010; Zablah et al., 2004). Furthermore, as value co-creation processes require the effective management of knowledge (Payne et al., 2008) the knowledge, skills and infrastructure to manage information must exist within a firm. Therefore, IT competency will positively influence the development of relational information processes within a firm. Thus:

> *H1* IT competency has a positive association with relational information processes.

Social media orientation

Culture represents a business-related CRM resource, one that Coltman (2007a) advocates is essential as a means of capitalising upon technological and human CRM resources. While many firms can employ similar technologies and possess similar skills, few will have a culture that understands how best to use these resources as a means of building superior customer relationships. Representative of the culture that exists within a firm (Noble, Sinha, & Kumar, 2002), a firm's orientation embodies the strategic dimension of CRM (Trainor, 2012). In order to implement CRM effectively, top management must demonstrate that CRM represents the strategic orientation of the firm (Becker, Greve, & Albers, 2009). As social CRM refers to the integration of social media technologies with traditional CRM processes and customer-facing activities (Trainor, 2012), a culture that understands the key role played by social networking sites in leveraging the co-creative competence of customers and adopts a strategic view of social CRM processes will be more successful in its implementation of relational information processes. Accordingly:

> *H2* Social media orientation has a positive association with relational information processes.

Performance outcomes of relational information processes

The process perspective of CRM, and the RBV within which it is embedded, conveys that it is resources, combined in a unique manner that facilitates the development of distinctive capabilities necessary to achieve superior levels of firm performance (Barney, 1991; Zablah et al., 2004). Similarly, the S-D logic details that it is through the identification and development of operant resources that superior firm performance is achieved (Madhavaram & Hunt, 2008). CRM literature indicates CRM initiatives positively impact financial (Fan & Ku, 2010) and non-financial aspects of firm performance (Jayachandran et al., 2005).

Customer performance

The effective management of customer relationships should result in higher levels of customer satisfaction (Jayachandran et al., 2005; Mithas et al., 2005), loyalty (Gustafsson & Johnson, 2005), retention (Jayachandran et al., 2005) and acquisition (Reinartz et al., 2004). The information acquired through customer interactions provides a means whereby customer needs and expectations can be understood and responded to more effectively, thus enhancing customer performance outcomes (Kim et al., 2010; Mithas et al., 2005). This is just as important in the context of social CRM, as the value co-creation potential offered by social networks should, through the increased interactivity they afford between a firm and its customer's, further enhance levels of customer performance through appealing to the active role desired by customers (Trainor, 2012). From a customer performance perspective, appealing to the active role desired by customers' results in enhanced satisfaction and loyalty (van Doorn et al., 2010). In addition, customer satisfaction and loyalty results in positive word of mouth, referrals, and the generation and dissemination of information which aids in the acquisition of new customers as well as the retention of existing ones (van Doorn et al., 2010). Therefore, the interaction and information management capabilities evident within relational information processes act as a means of understanding and responding to customer needs and expectations. When combined with the social capabilities offered by social networking sites, relational information processes, and hence customer performance, are further enhanced. Therefore:

H3 Relational information processes have a positive association with customer performance.

Financial performance

Given that CRM is concerned with the management of profitable customer relationships, the performance outcomes of CRM initiatives must communicate the financial outcomes of CRM initiatives (Boulding et al., 2005). The positive impact of CRM initiatives on financial aspects of performance has been observed (Fan & Ku, 2010) but continues to be of critical importance in light of the failure of many CRM initiatives (Ahearne, Rapp, Mariadoss, & Ganesan, 2012). As noted by Trainor (2012), capitalising on social CRMs ability to engage customers in the co-creation of value is, through the forging of stronger relationships with customers essential in order to positively impact financial performance. That is, the increased levels of customer engagement afforded by social technologies serves to enhance the bonds that exist between a firm and its customers, which in turn positively impacts financial performance (Stone & Woodcock, 2013). Given the depth and networked interactions facilitated by social networking sites, such outcomes and behaviours should be even more prevalent. Hence:

H4 Relational information processes have a positive association with financial performance.

Method

Sample and data collection

Given the high contact and experiential nature of the hospitality industry (Shaw et al., 2011), data were collected from the Irish hotel industry.

The conceptual framework was tested using data obtained from survey research. The research instrument took the form of a survey questionnaire. Respondents were offered the opportunity to complete a postal or an online version of the questionnaire, accessible via a link included in the cover letter accompanying the postal questionnaire. The sampling frame consisted of all hotels in the Republic of Ireland who have established a social networking presence. Based on extant literature, a social networking presence was defined as the development of an active profile and/or participation on social networking platforms. A census of hotels in the Republic of Ireland was obtained and vetted to establish how many were still in operation and had a social networking presence. Only hotels with an active social networking presence were considered for inclusion in the study. If hotels had not been active on social networking platforms in the previous three months, respondents were considered not to have maintained an active social networking presence. Of the 854 hotels in active operation in the Republic of Ireland, 757 (88.64%) had developed a profile on a social networking site. Consistent with previous research, the survey was mailed to those individuals listed as senior managers for each hotel (Haugland, Myrtveit, & Nygaard, 2007; Ordanini & Parasuraman, 2011). Follow-up telephone calls were made two weeks after mailing questionnaires. A sample size of 120 surveys was obtained, resulting in a response rate of 15.85%. Similar or smaller sample sizes have been obtained in previous CRM studies (e.g. Coltman 2007b; Kermati et al. 2010; Nguyen and Waring 2013 obtained sample sizes of 91, 77 and 126, respectively).

Measure development

The measures employed to test the conceptual framework are detailed in the appendix. Consistent with Jayachandran et al. (2005), relational information processes is conceptualised as a higher order construct. This also follows Coltman's (2007a) proposition of CRM as an organisational routine that is best conceptualised as a higher order construct.

Hotel size, star rating, location and *hotel type* represent covariates in the study. Studies demonstrate that each of these factors impact the use of IT within hotels, for example, smaller hotels, hotels with lower star ratings and hotels in rural locations tend to make smaller investments in information technologies (Matzler, Pechlaner, Abfalter, & Wolf, 2005). Hotel type (dependent or independent) may also impact the capabilities that can be developed by the hotel itself because being part of a dependent structure can limit the capabilities an individual hotel can cultivate (Haugland et al., 2007; Ordanini & Parasuraman, 2011). Hotel size, star rating, location and hotel type were measured using multiple-choice questions. As per previous studies, number of rooms was used as an indicator of hotel size (Ordanini & Parasuraman, 2011).

Data analysis approach

The survey data were analysed using the Statistical Package for the Social Sciences (version 20) for preliminary analysis and exploratory factor analysis. Partial least-squares structural equation modelling was then used to test the conceptual model. Partial least-squares structural equation modelling is increasingly being utilised in business and marketing research as it works well with complex models, smaller sample sizes and exhibits a greater degree of statistical power than co-variance-based structural equation modelling approaches (Hair, Hult, Ringle, & Sarstedt, 2014).

Results

A series of exploratory factory analyses were undertaken on the data. An iterative process was used to eliminate items with low loadings or cross loadings, leading to the elimination of a single item from the social media orientation and financial performance measure which demonstrated poor fit. Then, using partial least-squares structural equation modelling, the repeated indicators approach was employed to establish the hierarchical component model of the study. When employing the repeated indicators technique, the indicators of lower order constructs belonging to a higher order construct should be equal to avoid bias that emerges as a result of an unequal number of indicators per lower order construct (Hair et al., 2014). To have an equal number of indicators across the lower order constructs of the higher order construct in the study meant that some of the relational information process items had to be removed. Values greater than .70 demonstrate strong indicator reliability (Hair, Ringle, & Sarstedt, 2011). Therefore, as part of the removal process, those indicators with loadings lower than 0.7 were removed first. Where loadings were over .70, the lowest loadings were removed. The appendix lists the construct items retained after exploratory factor analysis and repeated indicators analysis.

Measurement model evaluation

Measurement model evaluation is concerned with establishing the reliability and validity of constructs. This involves evaluating four main criteria: indicator reliability, internal consistency, convergent validity and discriminant validity (Hair et al., 2014).

Indicator loadings were used to evaluate the *indicator reliability* of the model. Outer loading values exceeding .07 demonstrate strong indicator reliability (Hair et al., 2011). However, as detailed by Hair et al. (2014), values above .40 are acceptable subject to further investigation. The loadings of indicators on each lower and higher order construct are detailed in the appendix. Three loadings were lower than .70. However, these loadings exceeded the .40 value detailed by Hair et al. (2014) and were retained subject to validity tests.

Cronbach's alpha was used to assess *internal consistency*. Values above .70 are recommended in confirmatory research. Average variance extracted was employed to establish *convergent validity*, with values greater than .50 desirable (Hair et al., 2014). Cronbach's alpha and average variance extracted values for each of the constructs in the research study are presented in Table 1. As evident in Table 1, internal consistency and convergent validity are established.

Discriminant validity was assessed using the criteria proposed by Fornell and Larcker (1981). The average variance extracted for each construct was compared to the shared variance between each construct and the other constructs in the hypothesised framework. The results, as detailed in Table 2, demonstrate that the average variance extracted for each construct is greater than its shared variance with the other constructs, supporting discriminant validity.

Structural model evaluation

The structural model was first examined by assessing the coefficient of determination (R^2) and Stone-Geisser (Q^2) values. The coefficient of determination calculates the squared

Table 1. Internal consistency and convergent validity.

Construct	Cronbach's alpha (α)	Average variance extracted
IT competency	.858	.700
Social media orientation	.940	.679
Relational information processes	.948	.577
• Information reciprocity	.905	.777
• Information capture	.899	.767
• Information integration	.934	.836
• Information access	.934	.835
• Information use	.922	.810
• Value co-creation	.936	.838
Customer performance	.860	.603
Financial performance	.954	.756

correlation between the actual and predicted values of endogenous constructs in order to predict the accuracy of the model. The Stone-Geisser value utilises a blindfolding procedure to establish the difference between the true and predicted data points in order to predict the relevance of the model (Hair et al., 2014). Both the R^2 and Q^2 values act as a means of assessing the fit of partial least-squares structural equation models in the absence of global goodness-of-fit criteria (Hennig-Thurau, Houston, & Walsh, 2006; Hulland, 1999). Cohen (1988) details that R^2 values of .02, .15 and .35 represent small, medium and large effect sizes, respectively. Q^2 values exceeding 0 for endogenous constructs demonstrate good predictive relevance. Table 3 illustrates R^2 and Q^2 values, with results indicating that the model has a good predictive power.

The significance and relevance of the structural model relationships were then assessed. Direct effects were tested by running the partial least-squares algorithm. Bootstrapping was employed to obtain standard errors and t-values (Hair et al., 2014). *H1* posited that IT competency has a positive association with relational information processes. This association was supported ($\beta = .260$; t-value $= 3.783$; $p < .001$). *H2*, the positive association between social media orientation and relational information processes was also supported

Table 2. Discriminant validity.

Construct	IT competency	Social media orientation	Relational information processes	Customer performance	Financial performance
IT competency	**.700**	.100	.197	.104	.069
Social media orientation	.312	**.679**	.452	.080	.074
Relational information processes	.444	.672	**.577**	.086	.081
Customer performance	.322	.282	.294	**.603**	.445
Financial performance	.262	.272	.285	.667	**.756**

Notes: Diagonal entries represent the average variance extracted of each construct. Entries below diagonal are the correlations between constructs, and entries above diagonal are shared variance between pairs of constructs acquired from confirmatory factor analysis.

Table 3. Predictive accuracy and relevance of model.

Construct	R^2 value	Q^2 value
Relational information processes	.513	.237
• Information reciprocity	.274	.216
• Information capture	.552	.422
• Information integration	.676	.562
• Information access	.474	.399
• Information use	.787	.630
• Value co-creation	.701	.576
Customer performance	.107	.029
Financial performance	.119	.072

($\beta = .591$; t-value $= 10.569$; $p < .001$). Relational information processes were also found to have a positive association with customer and financial performance, supporting *H3* ($\beta = .281$; t-value $= 2.737$; $p < .01$) and *H4* ($\beta = .250$; t-value $= 2.576$; $p < .01$). The covariates did not have any significant effects on relational information processes and customer and financial performance.

Discussion and implications

In the era of the digitally empowered customer, there is an increasing amount of pressure on companies to involve customers in value creation processes as a means of building and maintaining customer relationships. Despite the emphasis of CRM literature on the co-creation of value, CRM initiatives have experienced mixed outcomes (Jayachandran et al., 2005; Zablah et al., 2004), with academics attributing these failures to the lack of interactivity provided by traditional CRM technologies (Trainor, 2012) and the dyadic rather than networked view of customer relationships (Gummesson, 2006). Despite the potential exhibited by social networking sites as platforms for value co-creation, literature is scant regarding how this can be undertaken in practice (Echeverri & Skålén, 2011; Payne et al., 2008). Hence, this study extends CRM literature in a social CRM context to investigate the set of activities carried out by a firm, inclusive of social networking activities, in order to co-create value with customers and as a result, build and maintain profitable, mutually beneficial customer relationships. Findings contribute significantly to social CRM literature by demonstrating that social networking sites have revolutionised the CRM process by allowing the true, co-creative nature of firm–customer relationships to be realised. Furthermore, the identification of relational information processes as those organisational routines necessary to co-create value with customers demonstrates that investment in CRM technology alone is not sufficient for improving firm performance (Coltman, 2007a; Trainor et al., 2014). Relational information processes must be designed around the co-creation of value, and social networking sites acting as enabler, provide a means whereby this can be achieved. The requisite human, technological and business resources must also exist within a firm, evident in the identification of IT competency and social media orientation as important antecedents to relational information processes.

Social CRM studies conducted by Choudhury and Harrigan (2014) and Trainor et al. (2014) also adopt aspects of relational information processes. However, this research study is the first to explicitly integrate social networking sites within each of these processes rather than as a separate social media technology use index. Accordingly, this research study fully investigates the impact of integrating these social CRM resources within

CRM processes. In addition, this study is the first to extend relational information processes to include a sixth dimension of value co-creation in light of the social capabilities afforded by social networking sites.

Theoretical implications

In addition to the significant contributions made to nascent social CRM literature, this research study has several theoretical implications, the first being the theoretical framework demonstrating how value co-creation can be managed in practice; an area in which extant value co-creation literature has been lacking (Echeverri & Skålén, 2011; Payne et al., 2008). Second, the theoretical framework is empirically tested and validated, which too has been lacking (Perks, Gruber, & Edvardsson, 2012). Consequently, the research study adds to the limited body of empirical research on how value co-creation processes can be employed in practice (Karpen et al., 2012; Payne et al., 2008). Third, interaction and information management capabilities, in line with the social capabilities afforded by social networking sites are identified as those capabilities necessary to engage customers in value co-creation practices. Inclusive of the antecedents and outcomes of these capabilities, this research contributes to what is outlined by Madhavaram and Hunt (2008) as an important research avenue in marketing strategy – the conceptualisation, measurement, antecedents and consequences of a firm's co-creation capability. Fourth, the identification of the central role played by social networking sites in value co-creation processes made by this study has also been lacking (Karpen et al., 2012; Payne et al., 2008). The integration of social networking sites with value co-creation processes makes a fifth contribution by extending CRM literature into a social CRM context (Greenberg, 2010; Trainor, 2012), demonstrating that social technologies provide the much needed interactivity and holistic view of relationships necessary to achieve the co-creation of value that has been emphasised by CRM literature.

Managerial implications

This research study has several managerial implications. First, value co-creation is occurring in practice and relational information processes act as key organisational routines if the co-creative potential of customers is to be leveraged effectively. Therefore, managers must focus on developing interaction and information management capabilities should they wish to pursue effective social CRM initiatives, which concentrate on enhancing customer relationships through the co-creation of value. The integration of social applications in these processes allows the true co-creative nature of the CRM process to be recognised.

Second, the identification of IT competency and social media orientation as key antecedents necessary to implement relational information processes successfully further aids managers on how value co-creation can be achieved in practice. The importance of IT competency reinforces that technology alone will not result in a competitive advantage. The skills and knowledge to use this technology must also exist within the firm. The importance of a social media orientation also demonstrates that a culture that understands the key role played by social networking sites in leveraging the co-creative competence of customers must exist. This must be reflected in the goals, policies and actions developed and implemented within the firm.

Third, the direct and positive impact of relational information processes on customer and financial performance demonstrates the strategic and marketing benefit of integrating social networking sites into organisational routines (Reibstein, Day, & Wind, 2009). Management must acknowledge the importance of social networking sites in enhancing customer relationships and integrate these sites within organisational routines as a means of engaging customers in those co-creative relationships that result in superior levels of firm performance.

Limitations and further research

Despite the numerous theoretical and managerial contributions made by the study, limitations do exist. The study focuses on role of social networking sites as platforms for value co-creation. While this does include a host of social media sites such as Facebook, Twitter, Google+ and Pinterest, the social media ecosystem encompasses a variety of platforms including blogs, forums and social bookmarking. Future research could focus on the role of other social media platforms in the value co-creation process.

Social networking sites were selected as the focus of the study due to the large growth experienced by these sites and the increased networking and interaction potential they offer on a social and professional level (Cheung & Lee, 2010). This could also be extended to include other digital innovations that have altered the manner in which customers interact with one another and how they produce, store and distribute information.

The study also focuses on a single industry, which, while improving internal validity, limits the generalizability of the research (Sheehan & Stabell, 2006). Consequently, the replication of the study in different industries provides an avenue for future research.

Disclosure statement

No potential conflict of interest was reported by the authors.

References

Ahearne, M., Rapp, A., Mariadoss, B. J., & Ganesan, S. (2012). Challenges of CRM implementation in business-to-business markets: A contingency perspective. *Journal of Personal Selling and Sales Management*, *32*(1), 117–130.

Ballantyne, D., & Varey, R. J. (2006). Creating value-in-use through marketing interaction: The exchange logic of relating, communicating and knowing. *Marketing Theory*, *6*(3), 335–348.

Barney, J. (1991). Firm resources and sustained competitive advantage. *Journal of Management*, *17*(1), 99–120.

Becker, J. U., Greve, G., & Albers, S. (2009). The impact of technological and organizational implementation of CRM on customer acquisition, maintenance, and retention. *International Journal of Research in Marketing*, *26*(3), 207–215.

Bharadwaj, A., El Sawy, O. A., Pavolu, P. A., & Venktraman, N. (2013). Digital business strategy: Toward a next generation of insights. *MIS Quarterly*, *37*(2), 471–482.

Boulding, W., Staelin, R., Ehret, M., & Johnston, W. J. (2005). A customer relationship management roadmap: What is known, potential pitfalls and where to go. *Journal of Marketing*, *69*(4), 155–166.

Brodie, R. J., Hollebeck, L. D., Juric, B., & Ilic, A. (2011). Customer engagement: Conceptual domain, fundamental propositions, and implications for research. *Journal of Service Research*, *14*(3), 252–271.

Cheung, C. M. K., & Lee, M. K. O. (2010). A theoretical model of intentional social action in online social networks. *Decision Support Systems*, *49*(1), 24–30.

Choudhury, M. M., & Harrigan, P. (2014). CRM to social CRM: The integration of new technologies into customer relationship management. *Journal of Strategic Management*, *22*(2), 149–176.

Cohen, J. (1988). *Statistical power analysis for the behavioral sciences*. Hillsdale, NJ: Lawrence Erlbaum Associates.

Coltman, T. (2007a). Can superior CRM capabilities improve performance in banking. *Journal of Financial Services Marketing*, *12*(2), 102–114.

Coltman, T. (2007b). Why build a customer relationship management capability? *Journal of Strategic Information Systems, 16*(3), 301–320.

Cova, B., & Salle, R. (2008). Marketing solutions in accordance with the S-D logic: Co-creating value with customer network actors. *Industrial Marketing Management, 37*(3), 270–277.

van Doorn, J., Lemon, K. N., Mittal, V., Nass, S., Pick, D., Pirner, P., & Verhoef, P. C. (2010). Customer engagement behavior: Theoretical foundations and research directions. *Journal of Service Research, 13*(3), 253–266.

Echeverri, P., & Skålén, P. (2011). Co-creation and co-destruction: A practice-theory based study of interactive value formation. *Marketing Theory, 11*(3), 351–373.

Eisenhardt, K. M., & Martin, J. A. (2000). Dynamic capabilities: What are they? *Strategic Management Journal, 21*(10/11), 1105–1121.

Ernst, H., Hoyer, W. D., Krafft, M., & Krieger, K. (2011). Customer relationship management and company performance – The mediating role of new product performance. *Journal of the Academy of Marketing Science, 39*(2), 290–306.

Fan, Y. W., & Ku, E. (2010). Customer focus, service process fit and customer relationship management profitability: The effect of knowledge sharing. *The Service Industries Journal, 30*(2), 203–223.

Fornell, C., & Larcker, D. F. (1981). Evaluating structural equation models with unobservable variables and measurement error. *Journal of Marketing Research, 18*(1), 39–50.

Fyrberg, A., & Jüriado, R. (2009). What about interaction? Networks and brands as integrators within service-dominant logic. *Journal of Service Management, 20*(4), 420–432.

Garrido-Moreno, A., & Padilla-Meléndez, A. (2011). Analyzing the impact of knowledge management on CRM success: The mediating effects of organizational factors. *International Journal of Information Management, 31*(5), 437–444.

Greenberg, P. (2010). *CRM at the speed of light: Social CRM strategies, tools, and techniques for engaging your customers* (4th ed.). New York, NY: McGraw-Hill.

Gummesson, E. (2006). Many-to-many marketing as grand theory. In R. F. Lusch & S. L. Vargo (Eds.), *The service-dominant logic of marketing: Dialog, debate and directions* (pp. 339–353). Armonk, NY: ME Sharpe.

Gummesson, E., & Mele, C. (2010). Marketing as value co-creation through network interaction and resource integration. *Journal of Business Market Management, 4*(4), 181–198.

Gustafsson, A., & Johnson, M. D. (2005). The effects of customer satisfaction, relationship commitment dimensions, and triggers on customer retention. *Journal of Marketing, 69*(4), 210–218.

Hair Jr, J. F., Hult, G. T. M., Ringle, G. M., & Sarstedt, M. (2014). *A primer on partial least squares structural equation modeling (PLS-SEM)*. Thousand Oaks, CA: Sage.

Hair, J. F., Ringle, C. M., & Sarstedt, M. (2011). PLS-SEM: Indeed a silver bullet. *The Journal of Marketing Theory and Practice, 19*(2), 139–152.

Haugland, S. A., Myrtveit, I., & Nygaard, A. (2007). Market orientation and performance in the service industry: A data envelopment analysis. *Journal of Business Research, 60*(11), 1191–1197.

Hennig-Thurau, T., Houston, M. B., & Walsh, G. (2006). The differing roles of success drivers across sequential channels: An application to the motion picture industry. *Journal of the Academy of Marketing Science, 34*(4), 559–575.

Hooley, G. J., Greenley, G. E., Cadogan, J. W., & Fahy, J. (2005). The performance impact of marketing resources. *Journal of Business Research, 58*(1), 18–27.

Hulland, J. (1999). Use of partial least squares (PLS) in strategic management research: A review of four recent studies. *Strategic Management Journal, 20*(2), 195–204.

Jayachandran, S., Sharma, S., Kaufman, P., & Raman, P. (2005). The role of relational information processes and technology use in customer relationship management. *The Journal of Marketing, 69*(4), 177–192.

Karpen, I. O., Bove, L. L., & Lukas, B. A. (2012). Linking service-dominant logic and strategic business practice: A conceptual model of a service-dominant orientation. *Journal of Service Research, 15*(1), 21–38.

Kermati, A., Mehrabi, H., & Mojir, N. (2010). A process-oriented perspective on customer relationship management and organizational performance: An empirical investigation. *Industrial Marketing Management, 39*(7), 1170–1185.

Kim, H. S., Kim, Y. G., & Park, C. W. (2010). Integration of firm's resource and capability to implement enterprise CRM: A case study of a retail bank in Korea. *Decision Support Systems, 48*(2), 313–322.

Lusch, R. F., & Vargo, S. L. (2006). Service-dominant logic: Reaction, reflections and refinements. *Marketing Theory*, *6*(3), 281–288.

Lusch, R. F., & Webster, F. E. (2011). A stakeholder-unifying, cocreation philosophy for marketing. *Journal of Macromarketing*, *31*(2), 129–134.

Madhavaram, S., & Hunt, S. D. (2008). The service-dominant logic and a hierarchy of operant resources: Developing masterful operant resources and implications for marketing strategy. *Journal of the Academy of Marketing Science*, *36*(1), 67–82.

Matsuno, K., Mentzer, J. T., & Rentz, J. O. (2005). A conceptual and empirical comparison of three market orientation scales. *Journal of Business Research*, *58*(1), 1–8.

Matzler, K., Pechlaner, H., Abfalter, D., & Wolf, M. (2005). Determinants of response to customer e-mail enquiries to hotel: Evidence from Austria. *Tourism Management*, *26*(2), 249–259.

Mithas, S., Krishnan, M. S., & Fornell, C. (2005). Why do customer relationship management applications affect customer satisfaction? *Journal of Marketing*, *69*(4), 201–209.

Nguyen, T. H., & Waring, T. S. (2013). The adoption of customer relationship management (CRM) technology in SMEs: An empirical study. *Journal of Small Business and Enterprise Development*, *20*(4), 824–848.

Noble, C. H., Sinha, R. K., & Kumar, A. (2002). Market orientation and alternative strategic orientations: A longitudinal assessment of performance implications. *The Journal of Marketing*, *66*(4), 25–39.

O'Cass, A., & Ngo, L. (2011). Examining the firm's value creation process: A managerial perspective of the firm's value offering strategy and performance. *British Journal of Management*, *22*(4), 646–671.

Ordanini, A., & Parasuraman, A. (2011). Service innovation viewed through a service-dominant logic lens: A conceptual framework and empirical analysis. *Journal of Service Research*, *14*(1), 3–23.

Payne, A., & Frow, P. (2005). A strategic framework for customer relationship management. *The Journal of Marketing*, *69*(4), 167–176.

Payne, A. F., Storbacka, K., & Frow, P. (2008). Managing the co-creation of value. *Journal of the Academy of Marketing Science*, *36*(1), 83–96.

Perks, H., Gruber, T., & Edvardsson, B. (2012). Co-creation in radical service innovation: A systematic analysis of microlevel processes. *Journal of Product Innovation Management*, *29*(6), 935–951.

Prahalad, C. K., & Ramaswamy, V. (2004). Co-creating unique value with customers. *Strategy & Leadership*, *32*(3), 4–9.

Raman, P., Wittman, M. C., & Rauseo, N. A. (2006). Leveraging CRM for sales: The role of organisational capabilities in successful CRM implementation. *Journal of Personal Selling and Sales Management*, *26*(1), 39–53.

Ramaswamy, V., & Gouillart, F. (2010). Building the co-creative enterprise. *Harvard Business Review*, *88*(10), 100–109.

Rapp, A., Trainor, K. J., & Agnihotri, R. (2010). Performance implications of customer-linking capabilities. Examining the complementary role of customer orientation and CRM technology. *Journal of Business Research*, *63*(11), 1229–1236.

Reibstein, D. J., Day, G., & Wind, J. C. (2009). Guest editorial: Is marketing academia losing it's way? *Journal of Marketing*, *73*(4), 1–3.

Reinartz, W., Krafft, M., & Hoyer, W. D. (2004). The customer relationship management process: Its measurement and impact on performance. *Journal of Marketing Management*, *41*(3), 293–305.

Shaw, G., Bailey, A., & Williams, A. (2011). Aspects of service-dominant logic and its implications for tourism management: Examples from the hotel industry. *Tourism Management*, *32*(2), 207–214.

Sheehan, N. T., & Stabell, C. B. (2006). Reputation and value creation in search shops. *The Service Industries Journal*, *26*(6), 597–613.

Sigala, M. (2009). E-service quality and web 2.0: Expanding quality models to include customer participation and inter-customer support. *The Service Industries Journal*, *29*(10), 1341–1358.

Susarla, A., Oh, J.-H., & Tan, Y. (2012). Social networks and the diffusion of user-generated content: Evidence from YouTube. *Information Systems Research*, *23*(1), 23–41.

Stone, M., & Woodcock, N. (2013). Social intelligence in customer engagement. *Journal of Strategic Marketing*, *21*(5), 394–401.

Teece, D. J., Pisano, G., & Shuen, A. (1997). Dynamic capabilities and strategic management. *Strategic Management Journal*, *18*(7), 509–533.

Tippins, M. J., & Sohi, R. S. (2003). IT competency and firm performance: Is organizational learning a missing link? *Strategic Management Journal*, *24*(8), 745–761.

Trainor, K. J. (2012). Relating social media technologies to performance: A capabilities-based perspective. *Journal of Personal Selling and Sales Management*, *32*(3), 317–331.

Trainor, K. J., Andzulis, J. M., Rapp, A., & Agnihotri, R. (2014). Social media technology usage and customer relationship performance: A capabilities-based examination of social CRM. *Journal of Business Research*, *67*(6), 1201–1208.

Vargo, S. L. (2008). Customer integration and value creation: Paradigmatic traps and perspectives. *Journal of Service Research*, *11*(2), 211–215.

Vargo, S. L., & Lusch, R. F. (2004). Evolving to a new dominant logic for marketing. *The Journal of Marketing*, *68*(1), 1–17.

Vargo, S. L., & Lusch, R. F. (2006). Service-dominant logic: What it is, what it is not, what it might be. In R. F. Lusch & S. L. Vargo (Eds.), *The service-dominant logic of marketing: Dialog, debate and directions* (pp. 43–56). Armonk, NY: M.E. Sharpe.

Vargo, S. L., Maglio, P. P., & Akaka, M. A. (2008). On value and value co-creation: A service systems and service logic perspective. *European Management Journal*, *26*(3), 145–152.

Wu, F., Mahajan, V., & Balasubranian, S. (2003). An analysis of E-business adoption and its impact on business performance. *Journal of the Academy of Marketing Science*, *31*(4), 425–447.

Zablah, A. R., Bellenger, D. N., & Johnston, W. J. (2004). An evaluation of divergent perspectives on customer relationship management: Towards a common understanding of an emerging phenomenon. *Industrial Marketing Management*, *33*(6), 475–489.

Appendix 1

	Loading	*t*-Value
IT *competency* (1: very low; 7: very high) (indicator source: Tippins & Sohi, 2003)		
Please rate the IT competency of your hotel relative to each of the following statements:		
• IT knowledge of relevant staff	.893	17.800
• IT skills of staff	.844	13.575
• Quality of IT infrastructure (such as hardware, software and support personnel)	.830	24.483
• Use of IT to manage market and customer information	.775	17.604
Social media orientation (1: strongly disagree; 7: strongly agree) (indicator source: newly developed measure)		
Our hotel …		
• believes that social media contributes significantly to the collection of important customer information	.724	14.612
• integrates social media with our overall marketing strategy	.878	32.344
• integrates social media with offline marketing channels	.870	38.209
• integrates social media with other online marketing channels	.888	37.220
• views social media as an important part of the marketing mix	.825	12.787
• has developed a social media marketing policies	.762	17.386
• has developed a social media marketing strategy	.777	20.664
• believes that using social media is integral to our overall company goals and strategy	.843	24.801
• intends to increasingly focus its marketing efforts on social media in the future	.832	25.547
Relational information processes (1: strongly disagree; 7: strongly agree)		
Information reciprocity (indicator *source*: Jayachandran et al., 2005)	.523	7.623
Our hotel uses social networking sites to …		
• enable our customers to have interactive communications with us	.930	64.397
• provide our customers with multiple ways to contact the organisation	.895	42.190
• focus on communicating periodically with our customers	.829	19.202
• maintain regular contact with our customers	.870	26.803
Information capture (indicator *source*: Jayachandran et al., 2005)	.743	18.822
Our hotel …		
• collects customer information from social networking sites on an ongoing basis	.908	42.259
• captures customer information relevant to social networking operations from internal sources within the organisation (such as sales, customer service and marketing staff)	.885	40.433
• uses social networking sites to collect customer information from external sources (such as market research agencies, syndicated data sources, and consultants' social networking site pages)	.832	24.657
• updates customer information collected from social networking sites in a timely fashion	.878	32.879
Information integration (indicator *source*: Jayachandran et al., 2005)	.822	26.869
Our hotel integrates customer information captured via social networking sites with …		
• customer information from the various functions that interact with customers (such as marketing, sales and customer service)	.883	26.784
• internal customer information	.920	58.128
• customer information from different communication channels (such as telephone, mail, email, the Internet, fax and personal contact)	.936	54.320
• information collected from various sources (such as functions and different communications channels) for each individual customer	.917	50.673
Information access (indicator source: Jayachandran et al., 2005)	.688	10.789
In our hotel, those employees who are responsible for communicating with customers via social networking sites …		

(Continued)

Appendix 1. Continued.

	Loading	t-Value
• find it easy to access required customer information	.925	52.425
• can access required customer information even when other departments/ functional areas have collected it	.909	36.345
• always have access to up-to-date customer information	.914	33.228
• have access to the information required to manage customer relationships	.908	31.176
Information use (*indicator source*: Jayachandran et al., 2005)	.887	40.477
The combination of information from social networking sites and other business-to-customer interactions is used by our hotel to ...		
• develop customer profiles	.888	38.720
• assess customer retention behaviour	.911	55.307
• identify appropriate channels to reach customers	.923	52.232
• customise our offers	.877	30.281
Value co-creation (*indicator source*: O'Cass & Ngo 2011)	.837	27.609
The combination of information from social networking sites and other business-to-customer interactions is used by our hotel to ...		
• interact with customers to design offerings that meet their needs	.894	28.729
• provide services for and in conjunction with customers	.932	54.328
• co-opt (encourage direct) customer involvement in providing services for them	.918	29.430
• provide customers with supporting systems (assistance and support they need) to help them get more value	.918	53.324
Customer performance (1: much worse; 7: much better) (indicator source: Hooley, Greenley, Cadogan, & Fahy, 2005; Jayachandran et al., 2005; Rapp et al., 2010)		
Over the past year, in relation to our business performance ...		
• levels of customer satisfaction have been ...	.778	6.844
• levels of customer loyalty have been ...	.719	4.460
Over the past year, relative to major competitors ...		
• levels of customer loyalty have been ...	.656	3.762
• the acquisition of new customers has been ...	.852	7.915
• the retention of existing customers has been ...	.857	8.185
Financial performance (1: much worse; 7: much better) (indicator source: Hooley et al., 2005; Matsuno, Mentzer, & Rentz, 2005; Wu, Mahajan, & Balasubranian, 2003)		
Over the past year, relative to major competitors		
• overall performance has been ...	.862	20.733
• sales volume achieved has been ...	.851	20.042
• market share has been ...	.897	31.564
• overall profits have been ...	.891	31.571
• profit margins have been ...	.893	37.376
• return on investment has been ...	.886	35.573
• return on sales has been ...	.819	13.343
• return on assets has been ...	.851	29.153

The value of social presence in mobile communications

Ji Hee Song[a] and Candice R. Hollenbeck[b]

[a]College of Business Administration, University of Seoul, Seoul, South Korea; [b]Department of Marketing, Terry College of Business, University of Georgia, Athens, GA, USA

Texting via mobile devices is used as a primary means for day-to-day communications among an increasing number of consumers and, as a result of this trend, more companies are engaging with consumers and addressing service complaints using social media platforms, such as Facebook. This study addresses the use of two-way mobile texting via Facebook to resolve service complaints with applications from social presence theory. Research shows that marketers' warm emotions are important in addressing service complaints, yet prior works mainly focus on the significance of human warmth in face-to-face contexts. Therefore, this study uses an experimental design to investigate the value of social presence in mobile texting as a means for providing service recovery. In triangulating the data, we use focus groups in confirmatory analysis. The findings show that social presence cues add human warmth to text messages with respect to two-way communication perceptions, control perceptions, responsiveness perceptions, satisfaction, attitudes, and repurchase intentions. The article concludes with a discussion of the importance of social presence cues in improving customers' experiences and overall satisfaction.

Today, mobile phones are an integral part of consumers' lifestyles, and this trend will only continue to increase (Chen & Chang, 2013; Pynta et al., 2014). For service providers, it is important to understand that consumers are migrating from PCs to mobile devices when communicating with firms (Fulgoni & Lipsman, 2014). Texting allows for timely, abbreviated interactivity; we broadly define mobile texting as messages consumers send to companies in the form of brief, electronic communications with the expectations of a reply. Texts sent through cellular devices use minimal character spaces, and various online platforms are becoming popular mediums for interactive texts. For example, in this study we examine texts sent via Facebook messages. The challenge for companies is to balance short, concise text replies while also conveying social presence, or the salience of another person in a mediated environment (Short, Williams, & Christie, 1976).

With the vast array of opportunities for consumers to text companies, marketers and service providers must gain a holistic understanding of optimizing text communications as more consumers use their mobile devices as a primary means for day-to-day communications (Sago, 2010). The mobile phone market has shown steady growth, and usage rates are largely driven by the rise of smartphones among consumers across all demographics. The basic phone market has declined dramatically, but this is offset by the increase in smartphone ownership as well as the increase in average spending per device (Harland, 2015). Indeed, texting is an important part of consumers' daily routines, and even car manufacturers are revolutionizing vehicles with hands-free texting capabilities (Bird, 2013). Texting capabilities allow consumers to streamline their communications with straightforward and concise messages while on the go. The popularity of high-end smartphones is largely driven by consumers' desire to be ubiquitously connected, and this includes being able to text anywhere at any time, even while driving (Bird, 2013). Studies indicate that texting will continue to increase and serve as an important means for service providers in communicating with consumers (Wu, 2005).

Much of the research on mobile communications has focused on one-way 'awareness-driven' messages (Shankar, Inman, Mantrala, Kelley, & Rizley, 2011), while scant attention has been paid to the use of mobile technology for two-way communications, specifically using social media platforms to address service complaints. Research estimates that approximately half of complaining customers are dissatisfied with firms' complaint-handling processes (Homburg & Fürst, 2005). More important, how marketers mechanistically respond to complaints directly affects customer satisfaction and loyalty (Homburg & Fürst, 2005). Research has shown that timely recovery methods decrease the likelihood of customer switching (Cambra-Fierro, Berbel-Pineda, Ruiz-Benítez, & Vázquez-Carrasco, 2013; Tsai & Su, 2009), and the ways a company responds to customer complaints, concerns, or queries are antecedents to loyalty and specific relational attachments (Chang & Hsiao, 2008; DeWitt, Nguyen, & Marshall, 2008). In terms of maintaining positive relationships with customers, the more empathic and intense an apology is, the more satisfied customers are, and, importantly, a late or inappropriate apology decreases consumer satisfaction (Roschk & Kaiser, 2013). It is clear that customers value the efforts companies invest in resolving problems, and if customers perceive a company as real and sincere in administering customer service, their satisfaction will increase even if a solution is not provided (Cambra-Fierro et al., 2013). In other words, it is important for companies to manage the perceptions of social presence by ensuring that customers are well aware of the sincere actions they are taking to resolve problems.

Social media is a practical context for examining social presence in mobile texting. With the rise in social media usage, companies cannot ignore platforms such as Facebook, Twitter, and Instagram to manage customer communications. Mobile is the largest growth area for digital advertising and will continue to be a key area for social media interactions in the next several years (Harland, 2015). Furthermore, social media is the most-used communication channel among various industries because it allows firms to build authentic relationships with customers (Harland, 2015). Customers want authentic and earnest resolutions to service failures. Indeed, empathy has the strongest impact on service recovery satisfaction, followed by intensity and timing (Roschk & Kaiser, 2013). Texting with consumers to resolve complaints provides timely responses; however, can communications via texting convey social presence, specifically empathy and human warmth? In this research, we aim to examine social presence cues in texting via Facebook communications between customers and firms to resolve service complaints. Research recommends that companies appeal to consumers' desire to interact with firms by offering more opportunities for them to voice problems so that issues can be quickly identified and resolved and negative word of mouth can be minimized (Chelminski & Coulter, 2011).

This study is anchored in social presence theory, which predicts that a receiver is more likely to understand the intended message when messages are rich in socially obvious connotations (Short et al., 1976). In short, mobile devices provide marketers with tools to interact with consumers in real time (Balasubramanian, Perterson, & Jarvenpaa, 2002), enabling service recovery as problems occur. However, companies must fully understand that while customers want abbreviated language and concise sentences, they also desire human warmth in the context of messages. Technology-assisted service failure recovery that relies on mobile devices for communications, without a holistic understanding of the value of social presence, may result in a perceived lack of interpersonal contact (Cowles & Crosby, 1990; Dabholkar & Spaid, 2012). Research has shown that verbal cues, body language, and emotional displays lead to greater satisfaction in face-to-face service encounters (Jiangang, Fan, & Feng, 2011) and that the emotional environment provided by the retailer can influence customer perceptions during service recovery (Harrison-Walker, 2012). Thus, because emotions play a critical role in resolving service complaints, social presence cues are important components to consider in text messages.

The remainder of this article proceeds as follows: We begin with a literature review of social presence theory and interactivity theory to provide a theoretical foundation demonstrating the importance of social presence in marketing communications. We build on these two theories to develop a conceptual framework that demonstrates the mediating effect of social presence cues in the context of service recovery. Next, we describe our experimental design, investigating different service-level situations in which participants must resolve a problem using mobile texting via Facebook. We also explain how focus groups helped further substantiate and validate our findings. Finally, we discuss the results by revealing how the incorporation of social presence cues in texting enhances perceived customer value.

Conceptual background and hypotheses

Social presence cues

Informational cues refer to the ways information can be communicated, such as text, verbal cues (e.g. tone of voice), or nonverbal cues (e.g. facial expression) (Daft & Lengel, 1986). Often, information cues are used to display acceptance, stimulate intimacy, emphasize important points, and, more generally, help receivers understand content more efficiently and effectively. Without the use of proper cues, it takes longer for receivers to fully understand and process messages (Williams, 1977). The lack of verbal and nonverbal cues can significantly affect social presence (Short et al., 1976). Therefore, a sender's ability to effectively transmit emotionally sensitive information (e.g. service problems) without the luxury of facial expressions, direction of looking, posture, dress, and vocal cues is often challenging. Yet effectively communicating emotionally sensitive information so that the receiver perceives the message as sincere and earnest is an indication of social presence and richness.

Social presence has two dimensions: (1) intimacy (interpersonal versus mediated) and (2) immediacy (asynchronous versus synchronous) (Short et al., 1976). Social presence theory predicts that communication media vary depending on their ability to create a sense of intimacy and immediacy. People tend to perceive face-to-face communication as more sociable because it carries more *intimacy* cues, such as eye contact, smiling, and friendly body language. For example, the use of video chat in MSN messenger triggers greater *immediacy* than the use of e-mail communication because it can transmit more information in real time with visual cues. Furthermore, research has found that videotext is a better medium than teletext in delivering sensitive information (Cowles & Crosby, 1990).

The level of social presence can vary for any given medium of communication (Short et al., 1976), and e-service providers have adopted numerous methods to enhance their level of social presence. Exemplar technologies include visual features such as virtual chatting, avatars representing general graphic information personified by technology (Holzwarth, Janiszewski, & Neumann, 2006), and the use of pictures and three-dimensional images in product description pages. Studies have empirically tested and highlighted the positive effects of visual features on consumers' perceptions and behaviors (Cui, Wang, & Xu, 2010; Dabholkar & Spaid, 2012; Holzwarth et al., 2006).

Indeed, social presence is important in managing customers' perceptions. However, scant attention has been paid to social presence cues in texting. Previous studies have predominately focused on 'social presence' in e-shopping (Cui et al., 2010; Holzwarth et al., 2006) or product description pages (Hassanein & Head, 2005). In addition, although research (Cui et al., 2010; Holzwarth et al., 2006) has investigated nonverbal cues with the design of communication interfaces (e.g. avatars and emoticons), scarce research has addressed the effects of verbal cues (e.g. welcoming consumers by their first name and addressing consumers with socially rich text conveying positive emotions) in two-way text communications.

In the current study, mobile texting provides a unique context in which nonverbal communication is limited. Compared with e-mail, texting consists of short, concise communications and, as such, must rely on verbal cues, such as the tone of text (i.e. personal tone) and the level of connection (i.e. use of 'I' instead of 'we'). Texting is widely popular among consumers because of its time-efficient nature, and though this can be a plus for customers, texting is one of the most challenging forms of communication for service providers in terms of conveying empathy and intimacy. Service providers must rely on well-crafted verbal cues to convey social presence and richness. This study is positioned on the idea that verbal cues play a major role in enhancing the effectiveness of service recovery communications, particularly in the context of texting, when the ability to communicate empathy and intimacy is significantly more challenging than other forms of communication. The goal of such texting is to provide timely service recovery while managing customers' perceptions.

Research has treated perceived interactivity as a key concept in computer-mediated communications (Song & Zinkhan, 2008). Perceived interactivity includes three dimensions: (1) perceived two-way communication, (2) perceived control, and (3) perceived responsiveness (Liu, 2003; McMillan & Hwang, 2002; Wu, 2005). Perceived two-way communication refers to interpersonal or face-to-face communications (Ha & James, 1998). Note that customers can use multiple verbal and nonverbal cues when engaging in face-to-face communications. Customers perceive communications as more reciprocal and interpersonal if social presence cues are embedded in the message. For example, consider a situation in which a customer texts an airline company via Facebook or Twitter about how dissatisfied she is about the loss of her baggage. A service recovery reply sent by a service representative identifying herself with a name (i.e. social presence cue) rather than a machine-generated message without any personal identification is likely to enhance the social presence and richness of the text message, and in turn the customer is likely to perceive the message as sincere and earnest. Thus, we propose the following:

H1a: Greater verbal social presence cues are associated with greater perceived value of two-way communications.

Perceived control describes the amount of control a consumer feels over the process or outcome of a service encounter (Yen, 2005). Perceived control is enhanced when consumers have more options in service encounters or the ability to predict the process of service. In particular, telepresence theory explains the concept of perceived control in computer-mediated communications (Steuer, 1992). According to this theory, 'a medium's structure (e.g. design, site map, and navigation) contributes to consumers' perceptions of telepresence' (i.e. experience of presence in an environment by means of a communication medium) (Steuer, 1992, p. 76) and subsequently influences their perceptions of control. Thus, when a text message facilitates social presence cues (medium's structure), consumers' confidence and perceptions of telepresence should be enhanced, and their predictions of the two-way communication outcomes should be more favorable. Similarly, a study of Internet self-service technology identifies perceived control as a significant attribute influencing quality satisfaction (Yen, 2005). Thus, consumers should perceive text communications with social presence cues as more controllable. Therefore, we propose the following:

> *H1b*: Greater verbal social presence cues are associated with greater perceived user control over the service outcome.

Perceived responsiveness reflects a user's sense of how responsive a communication medium as a system is to his or her actions (Wu, 2005). Often, perceived responsiveness is discussed in terms of communication speed (i.e. amount of time sending and receiving messages) (Kiousis, 2002). Therefore, real-time chatting is more responsive than an exchange of e-mails or documents. In service marketing, benefits such as saving time and effort will satisfy customers regardless of the technology implemented by marketers (Yen, 2005). However, consumers' perceived speed is influenced not only by objective speed (e.g. actual waiting time) but also by subjective speed. Subjective speed is influenced by various communication features such as message content (Song & Zinkhan, 2008) and website background color during downloading (Gorn, Chattopadhyay, Sengupta, & Tripathi, 2002). Thus, consumers perceive communications as faster when they receive personalized messages rather than messages with no personalization (Song & Zinkhan, 2008). Similarly, they may perceive mobile texting as faster and more responsive when text messages are embedded with social presence cues. Thus, we propose the following:

> *H1c*: Greater verbal social presence cues are associated with greater perceived value of responsiveness.

Understanding how social presence cues influence message effectiveness is important for service providers. Firms must manage and maintain strong customer relationships (e.g. integrated marketing, communications with customers, and customer support services) (Pan & Lee, 2003), and a consumer's computer-mediated communication experience with a firm is an antecedent of communication effectiveness measures, such as purchase intentions, satisfaction, loyalty, and word-of-mouth behavior (Moore & Moore, 2004; Strauss & Hill, 2001). Within the context of this study, in which consumers are complaining about service failure after a purchase, we assess communication effectiveness via mobile texting by asking the participants to rate the following three measures: (1) satisfaction, (2) attitude toward the firm, and (3) repurchase intention. Therefore, we propose the following:

> *H2*: Greater verbal social presence cues are associated with (a) greater perceived satisfaction with the overall texting experience, (b) more positive attitudes toward the firm, and (c) greater repurchase intentions.

Building on prior empirical research on perceived interactivity, we also expect that the three interactivity perceptions (i.e. two-way communication, control, and responsiveness) will mediate the relationship between social presence cues and communication effectiveness. Social presence theory proposes that adding human warmth positively affects interactivity perceptions and subsequently enhances satisfaction, attitude, and repurchase intention. In addition, prior research on interactive communications has found a positive relationship between interactivity perceptions (i.e. two-way communication, control, and responsiveness) and attitudes and behavioral intentions (i.e. purchase, repurchase, and revisit intentions) (Johnson, Bruner, & Kumar, 2006; Liu, 2003; McMillan & Hwang, 2002). Therefore, we propose a mediation hypothesis.

> *H3*: Interactivity perceptions mediate the positive influence of verbal social presence cues on consumers' satisfaction, attitudes, and repurchase intentions.

Social presence cue and service recovery

Short et al. (1976) argue that interactive and effective communications depend on (1) the degree of social presence associated with the communication and (2) the degree of social presence required by the communication. For example, when the nature of communication requires constant assessment of another person's reactions, the communication outcome is more likely to be influenced by the level of social presence cues. In contrast, the quality of communication is unaffected by the degree of social presence cues when the communication situation is clear and straightforward (e.g. simple information transmission). That is, as task/situation complexity and ambiguity increase, social presence has an increasing influence on communication outcomes. One key type of communication between firms and consumers is transaction-related messages (e.g. when customers expect actions from the firm), such as when customers contact an organization to complain about specific problems (e.g. related to delivery, service, and billing). In this type of situation, customers often expect immediate (or timely) feedback. For marketers, dealing with complaints due to service failure can be problematic because such complaints often involve complex and nonroutine communication activities (Cowles & Crosby, 1990).

Imagine that a customer sends a text message to complain about a billing error. The firm's response might not be personalized or prompt for many reasons, including (1) all available representatives are taking care of other customers, (2) the corresponding representative is not knowledgeable enough to resolve the issue and needs to transfer the case to another representative, or (3) it takes a long time to retrieve the customer's billing information and history. In this situation (low level of service recovery), including social presence cues in the exchange will create a more interactive and effective communication, regardless of whether the major concern is resolved. In contrast, if the firm sends a prompt response text message that recovers service failure (high level of service recovery), adding social presence cues will not make a significant difference in the communication outcome, including interactivity perceptions and communication effectiveness (i.e. satisfaction, attitude, and repurchase intention). That is, when the communication situation is clear (i.e. high level of service recovery), the positive effect of social presence cues is minimized (Short et al., 1976). Therefore, we predict the following:

> *H4a*: When the service recovery level is low (a firm is unable to resolve the problem promptly), *two-way communication* perceptions are greater for messages with verbal social cues than those without verbal social cues. Conversely, when the service recovery level is high, verbal social presence cues do not make a significant difference in *two-way communication* perceptions.

> *H4b*: When the service recovery level is low (a firm is unable to resolve the problem promptly), *control* perceptions are greater for messages with verbal social cues than those without verbal

social cues. Conversely, when the service recovery level is high, verbal social presence cues do not make a significant difference in *control* perceptions.

H4c: When the service recovery level is low (a firm is unable to resolve the problem promptly), *responsiveness* perceptions are greater for messages with verbal social cues than those without verbal social cues. Conversely, when the service recovery level is high, verbal social presence cues do not make a significant difference in *responsiveness* perceptions.

H5a: When the service recovery level is low (a firm is unable to resolve the problem promptly), *satisfaction* is greater for messages with verbal social cues than those without verbal social cues. Conversely, when the service recovery level is high, verbal social presence cues do not make a significant difference in *satisfaction*.

H5b: When the service recovery level is low (a firm is unable to resolve the problem promptly), *attitudes* are greater for messages with verbal social cues than those without verbal social cues. Conversely, when the service recovery level is high, verbal social presence cues do not make a significant difference in *attitudes*.

H5c: When the service recovery level is low (a firm is unable to resolve the problem promptly), *repurchase intentions* are greater for messages with verbal social cues than those without verbal social cues. Conversely, when the service recovery level is high, verbal social presence cues do not make a significant difference in *repurchase intentions*.

Experiment

Stimulus and participants

We tested *H4a–H4c* and *H5a–H5c* using a full factorial design. The experiment is a 2 (verbal social presence cues: with versus without) $\times$ 2 (service recovery level: high versus low) between-subjects factorial design. We established the level of social presence and service recovery level in four ways: low service recovery without social presence cue, low service recovery with social presence cue, high service recovery without social presence cue, and high service recovery with social presence cue. We conducted a virtual experiment using mobile phones. Participants were recruited from undergraduate and graduate business courses from a US public university. Students received class credit as an incentive for participation. Of the 133 participants, 53.4% were men and 46.6% women, and the mean age of all participants was 21.3 years (range: 19–47 years). The young adult demographic segment served as our sample pool because research suggests that Millennials on a global scale are early adopters of technological innovations and this generation prefers texting as their primary means of mobile communications (Chan-Olmsted, Rim, & Zerba, 2013; Chhateja & Jain, 2014; Thomas, 2014). In addition, research suggests that Millennials are more avid texters than other generations, and texting behavior has become so ubiquitous that it bleeds into their other daily activities (Olmsted & Terry, 2014). Millennials are so efficient and effective at communicating by text that a recent study claims that texting among young adults serves as a useful tool in managing relationships and potentially heightens relationship intimacy between two or more parties (McGee, 2014). Likewise, Millennials are the most devoted users of social media (Harland, 2015).

Procedures

Participants were instructed to read the following scenario:

> About a week ago, you purchased a t-shirt as a birthday gift for your best friend at E-store.com. Yesterday, you received the item and discovered that the store sent the wrong t-shirt.

Now you want to figure out how you can receive the correct item as soon as possible and how to return the wrong item. Using your smartphone, you send a text to E-store.com via Facebook.

After reading the scenario, participants were asked to send a text message to E-store. Participants then waited until they received a response text message from the store. Then, we collected dependent measures, including two-way communication perceptions, control perceptions, responsiveness perceptions, satisfaction, attitudes, and repurchase intentions. Last, participants were asked to complete survey questions on their demographic information, which concluded the experiment. We measured all items with a seven-point Likert-scale. Table 1 provides a description of the measurement approaches and Cronbach's alpha associated with each scale.

Independent variable manipulation

We manipulated the social presence cues in terms of intimacy. Text messages that include authentic names and the use of 'I' are considered more intimate. In addition, the use of a conversational style rather than a formal, detached style adds social presence (e.g. 'Hi Sam, I will be right with you' versus 'Support is unavailable at this moment'). Service recovery level is the extent to which the replying text message resolves the sender's complaints. In the context of this study, consumers are complaining about *not* receiving what they ordered. Therefore, under high service recovery conditions, consumers receive prompt instructions on how to get the right item and how to return the wrong item. Under low service recovery conditions, consumers' complaints are not promptly addressed. Detailed scenario manipulations are as follows (underlined text indicates the social presence cues in the message):

Low service recovery without social presence cues: This is E-store.com. Do you need assistance?

(customers are asking questions)

You will receive a response to your inquiry soon.

Low service recovery with social presence cues: Hi Sam, thank you for contacting E-store.com. I'm Anna, E-store's Online Assistant. How can I help you today?

(customers are asking questions)

Thank you for contacting me. I will personally look into your question and will text back soon.

High service recovery without social presence cues: This is E-store.com. Do you need assistance?

(customers are asking questions)

Text back with the address you would like the item to be shipped. You should receive the new item within 7–10 business days. You will also receive a return label.

High service recovery with social presence cues: Hi Sam, welcome to E-store.com. I'm Anna, E-store's Online Assistant. How can I help you today?

(customers are asking questions)

I am so sorry about this inconvenience. I will personally make sure that you receive the right item this time. Please text me with the address where you would like me to send the item. I will reship the item to you today and you will receive it within 7–10 business days. I will include a return label. Can I help you with anything else?

Table 1. Measures of dependent variables.

Construct	Measures
Two-way communication perception ($\alpha = .915$) (Liu, 2003; McMillan & Hwang, 2002; Wu, 2005)	1) This E-store.com facilitates two-way communication 2) The E-store.com gives me the opportunity to talk back 3) The E-store.com facilitates concurrent communication 4) The E-store.com enables conversation 5) The E-store.com does not encourage visitors to talk back (R) 6) The E-store.com is effective in gathering visitors' feedback
Control perception ($\alpha = .824$) (Liu, 2003; McMillan & Hwang, 2002; Wu, 2005)	1) While I was at the E-store.com I was always aware where I was 2) While I was at the E-store.com, I always knew what I was doing 3) While I was at the E-store.com, I was always able to say what I wanted to say 4) I was delighted to be able to choose what I could do 5) I feel that I have a great deal of control over my experience with E-store.com 6) The E-store.com is not manageable (R) 7) While I was at the E-store.com, I could choose freely what I wanted to say 8) While I was at the E-store.com, I had absolutely no control (R) 9) While I was at the E-store.com, my actions decided the kind of experiences I got
Responsiveness perception ($\alpha = .871$) (Liu, 2003; McMillan & Hwang, 2002; Wu, 2005)	1) The E-store processed my input very quickly 2) Getting information from the E-store is very fast 3) I was able to obtain the information I want without any delay 4) I felt I was getting instantaneous information from the E-store 5) The E-store is very slow in responding to my request (R) 6) The E-store answers my question immediately
Attitude toward the site ($\alpha = .948$) (Coyle & Thorson, 2001)*	1) Good/bad 2) Favorable/unfavorable 3) Like/dislike
Satisfaction ($\alpha = .795$) (Fornell, Johnson, Anderson, Cha, & Bryant, 1996)*	1) I am satisfied with the experience 2) This experience is exactly what I needed 3) The experience has not worked out as well as I thought it would (R)
Repurchase intention ($\alpha = .931$) (Zeithaml, Berry, & Parasuraman, 1996)*	1) I would consider this store for my future online shopping 2) The next time I purchase a t-shirt, I will buy this from this store 3) I would be willing to purchase from this store again

Note: R = reversed coding.
*Seven-point Likert-scale.

Pretest

In a pretest, we asked the participants to answer the following seven-point Likert-scale questions to check the manipulations for the level of service recovery: 'E-store resolved my problem very well' and 'What do you think about the level of service recovery of the E-store.com?' Fifty-five students participated in the pretest. The results showed significant differences in participants' perceptions of service recovery levels ($t = -11.407$, $df = 53$, $p < .05$). Next, we asked 20 participants the 4 manipulation questions to test for perceived social presence (Short et al., 1976). We used a seven-point Likert-scale to assess participants' perceptions of social presence using the following dimensions: impersonal/personal, unsocial/social, insensitive/sensitive, and cold/warm. Participants showed significant differences in their perceptions of social presence levels ($t = -2.895$, $df = 18$, $p < .05$).

In addition, we asked participants what communication method they preferred for most service complaint situations. We provided four options: (1) face-to-face contact with the firm, (2) calling the firm by telephone, (3) e-mailing the firm from a personal computer, and (4) texting the firm using a mobile device. Participants then answered a series of open-ended questions. Of the participants, 96% indicated texting as their first choice of communication due to user control and time convenience. This finding indicates that texting is a viable option when considering intervention strategies for service failure situations. Furthermore, in contrast with prior work, we did not find cognitive effort a barrier to texting with firms (Kleijnen, De Ruyer, & Wetzels, 2007). Cognitive effort refers to the complexity of the innovation and the effort associated with understanding and using the technology. All the participants considered texting an easy and convenient means of communications.

Results

The effect of social presence cues

We ran multivariate analysis of variance (MANOVA) to test *H1* and *H2*. MANOVA is more appropriate than a series of univariate analysis of variance when there is a possibility that the composite of the dependent variables explains an overall group difference (Hair, Black, Babin, & Anderson, 2009). First, multivariate difference measures (i.e. Pillai's trace, Hotelling's trace, Wilks's lambda, and Roy's largest root) are significant ($p < .05$), indicating that combined dependent variables vary across the different levels of social presence. Table 2 shows the summary result of the MANOVA, and Table 3 presents means and

Table 2. MANOVA result: main effects and interaction effects.

Variable	df	Verbal social presence F value (effect size)	Service recovery level F value (effect size)	Verbal social presence × service recovery level (Interaction Effect) F value (effect size)
Two-way communication perception	1	18.693* (.126)	19.166* (.129)	16.394 (.113)*
Control perception	1	12.271* (.087)	4.239* (.032)	3.937 (.030)*
Responsiveness perception	1	7.832* (.057)	12.901* (.091)	1.616 (.012)
Attitude	1	16.013* (.110)	13.951* (.098)	2.555 (.019)
Satisfaction	1	15.808* (.109)	30.739* (.192)	8.121 (.059)*
Repurchase intention	1	11.508* (.082)	8.804* (.064)	1.726 (.013)

*Significant at the 5% level.

Table 3. Means and confidence intervals.

| Independent variables | Verbal social presence cue | | Service recovery level | | Verbal social presence cue × service recovery level | | | |
| | | | | | Service recovery level-low | | Service recovery level-high | |
Dependent variables	Without social presence cue ($n=67$)	With social presence cue ($n=66$)	Low ($n=59$)	High ($n=74$)	Without social cue ($n=31$)	With social cue ($n=28$)	Without social cue ($n=36$)	With social cue ($n=38$)
Two-way communication perception	4.757 (4.485–5.029)	5.604 (5.327–5.880)	4.751 (4.461–5.040)	5.610 (5.351–5.868)	3.930 (3.531–4.329)	5.571 (5.152–5.991)	5.583 (5.213–5.954)	5.636 (5.276–5.996)
Control perception	4.542 (4.303–4.781)	5.146 (4.903–5.390)	4.667 (4.412–4.921)	5.022 (4.795–5.249)	4.194 (3.843–4.544)	5.140 (4.771–5.509)	4.891 (4.565–5.216)	5.153 (4.836–5.470)
Responsiveness perception	3.947 (3.652–4.242)	4.542 (4.242–4.842)	3.863 (3.549–4.177)	4.626 (4.346–4.906)	3.430 (2.998–3.863)	4.295 (3.840–4.750)	4.464 (4.063–4.865)	4.789 (4.398–5.179)
Attitude	3.906 (3.630–4.181)	4.700 (4.420–4.980)	3.932 (3.639–4.225)	4.674 (4.412–4.935)	3.376 (2.973–3.780)	4.488 (4.063–4.913)	4.435 (4.060–4.810)	4.912 (4.548–5.277)
Satisfaction	3.739 (3.462–4.015)	4.530 (4.249–4.811)	3.538 (3.289–3.876)	4.686 (4.424–4.948)	2.903 (2.498–3.308)	4.262 (3.836–4.688)	4.574 (4.198–4.950)	4.798 (4.433–5.164)
Repurchase intention	3.489 (3.200–3.778)	4.193 (3.902–4.490)	3.533 (3.226–3.841)	4.152 (3.877–4.426)	3.304 (2.619–3.467)	4.024 (3.578–4.470)	3.935 (3.542–4.329)	4.368 (3.985–4.751)

Note: The 95% confidence interval is in parentheses.

confidence intervals of each condition. The results provide support for *H1a*, *H1b*, and *H1c*. The participants in the social presence conditions perceived the text message as more two-way communication oriented ($M_{social} = 5.604$, $M_{nosocial} = 4.757$; $F(1, 129) = 18.693$, $p < .05$), controllable ($M_{social} = 5.146$, $M_{nosocial} = 4.542$; $F(1, 129) = 12.271$, $p < .05$), and responsive ($M_{social} = 4.542$, $M_{nosocial} = 3.947$; $F(1, 129) = 7.832$, $p < .05$) than those in conditions with no social presence cues. Similarly, the effects of social presence cues on satisfaction ($F(1, 129) = 15.808$, $p < .05$), attitudes ($F(1, 129) = 16.013$, $p < .05$), and repurchase intentions ($F(1, 129) = 11.508$, $p < .05$) were all significant, indicating that a text message with verbal social presence cues influences these communication effectiveness measures. Therefore, *H2a*, *H2b*, and *H2c* are supported. The results of *H1* and *H2* reveal that the use of social presence cues in communication messages increases interactivity and communication effectiveness. The effect sizes of social presence cues on interactivity perception and communication effectiveness were between .057 and .126. Social presence cues had the greatest effect on two-way communication perceptions (.126), followed by attitude (.110) and satisfaction (.109). Thus, social presence cues in text messages play a significant role in stimulating reciprocal communication, creating a positive image for the firm, and enhancing overall customer satisfaction. Finally, social presence cues were relatively weak determinants of responsiveness perceptions when compared with the other two interactivity perceptions (effect size = .057).

The mediating role of interactivity perceptions

We conducted Bootstrap estimation (Preacher & Hayes, 2004; Zhao, Lynch, & Chen, 2010) with 5000 resample as well as Sobel tests, to test the possible mediating effects of the three interactivity perceptions, with social presence cues as an independent factor, for the three site effectiveness measures. Table 4 provides the indirect effect sizes of social presence cues on site effectiveness, Sobel test Z-value, and confidence intervals. The results show that (1) two-way communication mediated the influence of social presence cues on attitudes and satisfaction; (2) control mediated the influence of social presence cues on attitudes, satisfaction, and repurchase intentions; and (3) responsiveness mediated the influence of social presence cues on attitudes, satisfaction, and repurchase intentions. Among the three mediators, responsiveness was the strongest (see the effect sizes in Table 4). The results also indicate that two-way communications do not significantly mediate the influence of social presence cues on repurchase intentions (see the Z-value and confidence interval,

Table 4. Mediation test results.

Dependent variables Mediators	Indirect effects of social presence through mediators (Sobel test: Z-value) (Bootstrap test: 95% CI)		
	Attitude	Satisfaction	Repurchase
Two-way communication perception	0.1695* (2.2725) (.0357–.3868)	0.1799* (2.4702) (.0441–.4028)	0.0952 (1.2807) (−.0321–.3078)
Control perception	0.1757* (2.4041) (.0610–.3490)	0.2217* (2.7282) (.0830–.4384)	0.1474* (1.9697) (.0224–.3607)
Responsiveness perception	0.2223* (2.4197) (.0673–.4278)	0.2774* (2.5359) (.0908–.5264)	0.2194* (2.3164) (.0599–.4634)

*Significant at the 5% level.
Note: If the confidence interval does not include 0, the indirect effect is significant.

which includes 0). However, the same bootstrap analysis with two-way communication as the only mediator confirmed that two-way communication is a significant mediator in the relationship between social presence cues and repurchase intentions ($M = .3382$, SE $= .11$, confidence interval $= .136, .616$; Sobel test: $Z = 3.044$, $p < .05$). Overall, the findings provide evidence for the mediating role of interactivity perceptions. Social presence cues positively influence communication effectiveness (i.e. satisfaction, attitudes, and repurchase intentions), and this effect is mediated by the three interactivity perceptions, stimulated by social presence cues. Therefore, *H3* is supported.

The significant main effect of social presence cues leads to the following question: Does increasing the level of social presence always guarantee higher levels of interactivity and communication effectiveness? Under which situations is the value of social cues maximized in new media communications, such as mobile texting? In other words, when should firms use social cues in mobile texting? To answer these questions, we examine the interaction effect between social presence and the level of service recovery.

The interaction effect: social presence cues and service recovery level

H4 and *H5* predict the interaction effect of social presence cues and the service recovery level. We conducted a MANOVA to test both *H4* and *H5*. The multivariate difference measures (i.e. Pillai's trace, Hotelling's trace, Wilks's lambda, and Roy's largest root) were significant ($p < .05$), indicating that the combined dependent variables varied across the different levels of social presence and service recovery. The interaction effect of a social presence cue and service recovery levels on two-way communication (*H4a*) was significant ($F(1, 129) = 16.394$, $p < .05$) (see Figure 1 and Table 2). That is, when the firm transmitted text messages that did not resolve customers' service failure issue (low service recovery), their two-way communication perceptions were greater for the message with social cues ($M = 5.571$) than the message without social cues ($M = 3.93$; F $(1, 129) = 31.424$, $p < .05$). Conversely, when the text message solved customers' problem (high service recovery), the impact of social presence cues on their two-way communication perceptions was not significant ($M_{highservicerecovery \times socialcues} = 5.636$, $M_{highservicerecovery \times nosocialcues} = 5.583$; $F(1, 129) = .04$, n.s.). Similarly, the findings show significant interaction effects on control perceptions ($F(1, 129) = 3.937$, $p < .05$) (*H4b*). In the case of high service recovery, the effect of social presence cues on control was not significant ($M_{highservicerecovery \times socialcues} = 5.153$ $M_{highservicerecovery \times nosocialcues} = 4.891$; $F(1, 129) = 1.30$, n.s.). Adding social presence cues makes a significant difference in control perceptions when the service recovery level is low ($M_{lowservicerecovery \times socialcues} = 5.14$,

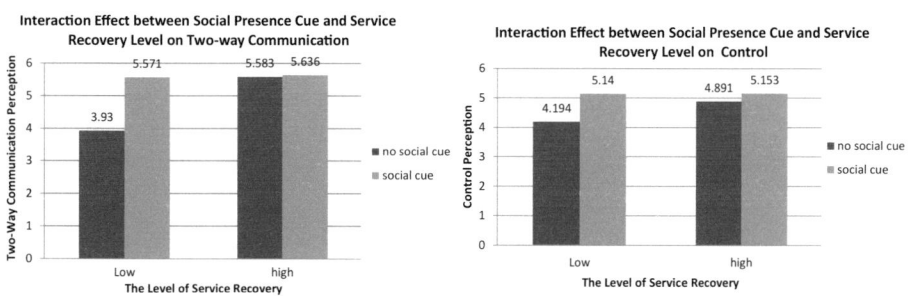

Figure 1. Interaction effect on two-way communication and control perceptions.

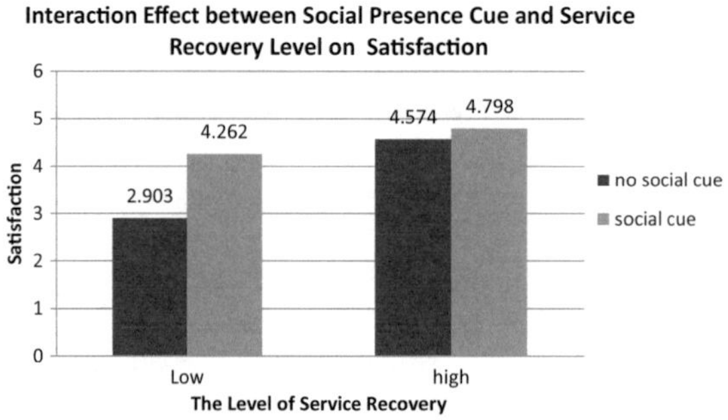

Figure 2. Interaction effect on satisfaction.

$M_{lowservicerecovery \times nosocialcues} = 4.194; F(1, 129) = 13.518, p < .05$). Interaction effects on consumers' perceptions of responsiveness were not significant ($F(1, 129) = 1.616, p > .05$) (*H4c*). Therefore, *H4* is partially supported.

The interaction effects of social presence cues and service recovery levels on satisfaction were significant ($F(1, 129) = 8.121, p < .05$) (see Figure 2 and Table 2). As *H5a* predicted, when the service recovery was low, adding social cues was a significant predictor of satisfaction ($M_{lowservicerecovery \times socialcues} = 4.262, M_{lowservicerecoverynosocialcues} = 2.903$; $F(1, 129) = 20.917, p < .05$). However, under the high service recovery condition, adding social presence cues did not have a significant impact on satisfaction ($M_{highservicerecovery \times socialcues} = 4.798, M_{highservicerecovery \times nosocialcues} = 4.574$; $F(1, 129) = .716$; n.s.). Although the interaction effects on attitudes (*H5b*) and repurchase intentions (*H5c*) are not significant, similar patterns were found for both (see Table 3). For example, under the low service recovery condition, attitudes of consumers who received a message with social presence cues ($M = 4.488$) were higher than those of consumers who received a message with no social presence cues ($M = 3.376$).

Whereas repurchase intentions and attitudes are more associated with the overall evaluation of the firm, satisfaction was connected with a specific experience (i.e. communication experience) with the firm. Therefore, under low service recovery conditions (i.e. service failure is not resolved), adding social cues to communication messages is not enough to significantly enhance overall evaluations and images (i.e. attitudes and repurchase intentions) of the firm. Consumers expect more tangible gains from marketers, and simply including social cues is not enough to increase repurchase intentions and attitudes. However, social presence cues play an important role in consumers' specific experiences with the firm (i.e. communication messages), even when service failure is not resolved.

Validation

Our findings show that the inclusion of social presence cues in texting enhances consumers' perceptions of interactivity, satisfaction, attitudes, and repurchase intentions. In addition, the level of service failure moderates these relationships. To further substantiate and validate the results, we gathered additional qualitative insights on consumers' attitudes toward mobile interactions with firms and a firm's general use of social media, specifically Facebook, in the context of service failure.

We conducted three focus groups with consumers who currently own smartphones and use them on a weekly basis to interact with retail firms. All participants had shopped for and purchased items using their smartphone within one month before the focus group, and all used Facebook on a daily basis. Thirty-six participants were recruited from undergraduate business courses at a large US public university and equally divided into 3 groups. Participants were 16 men and 20 women, ages ranged from 20 to 26 years, and all were from diverse ethnic backgrounds: Caucasian, African American, Asian, and Hispanic.

The focus groups were useful in validating consumers' attitudes toward using mobile devices to connect with firms and the potential costs and benefits of using texts via Facebook messages to interact with firms, specifically in the context of service failure. The focus groups began with a warm-up discussion about the advantages and disadvantages of mobile devices in interacting with firms. Then, participants were asked to complete three projective exercises (sentence completion, storytelling, and picture drawing) intended to delve into unconscious beliefs and attitudes toward mobile communications.

The projective exercises revealed that consumers value social presence cues in mobile texting, while also valuing the time-related benefits of short, concise messages. Participants told stories about their most valued means of communications, which involved thoughtful details that made them feel special. Participants also told stories about their worst texting experiences; these stories revolved around similar issues: miscommunication, misunderstanding, impersonal responses, and 'too abbreviated' responses.

The participants were then asked to act as if they were the head of the customer service department for a large retail firm and, with this position of authority, to describe how they would manage service failure situations using mobile texting; specifically, how would they emphasize the benefits of mobile communications to customers, and what kinds of messages would be standard communications. Finally, participants provided service failure scenarios and responded to customers' complaints using mobile texting. Participants provided two separate responses: one response with social presence cues and one response without cues. Toward the end of the focus group, participants discussed the possible outcomes when using each response (with and without social cues).

Our analysis of verbatim focus group transcripts involved an iterative, part-to-whole strategy from which we aimed to develop a holistic understanding of the value of mobile texting to resolve service complaints. Aided by Atlas software and following the data analysis and interpretation guidelines outlined by prior research (Spiggle, 1994), we utilized our theoretical background to explain behavior, such as social presence theory, interactivity theory, double deviation, and asymmetric disconfirmation theory. The thematic categories that emerged from data analysis included the value of social presence cues in texting, perceived user control over the outcome during service failure situations, perceived firm responsiveness when service fails, and sources of satisfaction and dissatisfaction when texting to resolve service failure. Table 5 provides exemplary quotes and illustrates data categorization. In particular, participants experienced empathy and intimacy when mobile texts embodied cues of social presence; this theme supports social presence theory and interactive theory. In addition, in the case of service failure, participants expected inclusions of human-like service recovery efforts, especially when service recovery was not successful (e.g. social presence theory). Therefore, with no contradictory data, the focus groups provided confirmatory data corroborating our proposed theories.

Table 5. Exemplary quotes from focus group data.

Thematic categories	Discussion context	Exemplary quotes from focus groups	Literature comparisons (related hypotheses)
Value of social presence cues	The benefits of mobile texting for service recovery	'Texting is great because it is a great way to connect with companies especially when I need to get a problem solved. It saves time and effort for customers … and when the company responds with legitimate, real interest [human warmth] in my problem, it just boosts my loyalty.' (Caucasian, F, 25)	Social presence theory (*H4*, *H5*)
	Descriptions of favorable mobile texting experiences	'It makes me feel special when someone uses my name in a text or relays an immediate sense of concern.' (African American, F, 21)	Social presence theory (*H1*)
Perceived user control	Texting for the purpose of service recovery	'When somebody or some physical person is actually responding to my texts, I know that I'm gonna get the problem fixed.' (Hispanic, M, 24)	Interactivity theory (*H1*)
Perceived firm responsiveness	Examples of good communication via texting during service failure	'There is something about having the reassurance that someone is helping you … that personal dialogue makes you believe in the company and you know that individual customers are important for that company.' (Caucasian, F, 21)	Interactivity theory (*H1*)
Sources of dissatisfaction	Examples of bad communication via texting during service failure	'There is so much miscommunication that happens when texting because I can't see the other person's face or reaction to my text and the lack of personalization just exaggerates the miscommunication.' (Caucasian, F, 22)	Social presence theory (*H4*, *H5*)
	Descriptions of worst mobile texting experiences	'The worst texting experiences are when I don't feel like I'm talking with another person … I hate computer generated responses!' (Caucasian, M, 22)	Social presence theory (*H2*)
	Reflections on service failure without human warmth	'This [texts without human warmth] would make me feel like a number – not a person.' (Caucasian, F, 21)	Interactivity theory (*H1*)

Discussion

This study contributes to the marketing and service literature in several ways. First, we show that mobile texting is a viable means for addressing service complaints via social media platforms such as Facebook. Much of the literature to date has focused on marketers' use of mobile technologies and social media tools as one-way promotional and awareness campaigns. Moreover, mobile technologies are considered ineffective in resolving situations of negative attribution by customers (Dabholkar & Spaid, 2012). Here, we demonstrate that mobile texting via Facebook can be useful in resolving service complaints in a timely and efficient manner using two-way dialogue. Although the primary advantage of mobile technology is time convenience (Kleijnen et al., 2007), our findings highlight the benefits of adding social presence cues to text messages, particularly in the context of service failure. In a culture that thrives on keeping text messages short with abbreviated sentences, we show that incorporating social presence cues in two-way texting is important in cultivating empathy and human warmth.

Second, the findings contribute to the understanding of mobile technology and its perceived value as a communication channel. According to Kleijnen et al. (2007), cognitive effort and perceived risk are antecedents of perceived value. However, our focus group participants indicated that *cognitive effort* is not applicable to younger demographics because of the pervasiveness of mobile communications. The participants in the pretest conveyed a preference for mobile texting over other communication channels (e.g. telephone and face-to-face). This finding is consistent with research that suggests that young adults are leading the way in mobile technological innovations (Preacher & Hayes, 2004). Regarding *perceived risk*, research defines this concept as a cost weighed against benefits in value perceptions (Kleijnen et al., 2007). With this definition in mind, our findings suggest that perceived risks can be assuaged by embedding social presence cues in text messages. More specifically, our findings show that the inclusion of social presence cues in texting enhances consumers' perceptions of interactivity in terms of two-way communication, control, and responsiveness (*H1a*, *H1b*, and *H1c*). Furthermore, the incorporation of social presence cues provides more effective message communications, and this, in turn, increases consumers' satisfaction, attitudes, and repurchase intentions (*H2a*, *H2b*, and *H2c*). Our findings reveal that with the inclusion of social presence cues, mobile texting can be an effective form of communications in resolving service failures.

Third, our findings are concurrent with prior work suggesting that empathy and time convenience are most important when using mobile texting (Kleijnen et al., 2007). On the one hand, our participants use mobile texting for time convenience; on the other hand, they prefer lengthier messages in the context of service failure. In balancing time convenience with empathy, it is important to incorporate social presence (when compared with shorter messages without social presence cues). Using technology-based communications without a clear understanding of social presence could lead to negative perceptions among customers. Therefore, companies need a thoughtful, planned process for service recovery. In essence, text messages should be concise but embedded with social cues.

Fourth, during service recovery situations in which the remedy is delayed, we find that the value of social presence cues is heightened. The effect of social presence cues on two-way communication perceptions, control perceptions, and satisfaction is influenced by the degree of social presence required for a given situation (i.e. the level of service recovery) (*H4a*, *H4b*, and *H5a*). When a firm is unable to resolve a problem promptly, embedding social cues in text messages can improve the virtual mood. This is analogous to the way marketers' positive emotions can minimize customers' negative emotions in face-to-face contexts (Jiangang et al., 2011).

Fifth, the nonsignificant interaction effect on responsiveness perceptions (*H4c*) indicates that under low service recovery conditions, customers still perceive a text message as more reciprocal and controllable when social presence cues are evident. However, customers might not perceive the message as responsive (i.e. fast and immediate) because adding social presence cues is not enough to increase their perceptions of responsiveness when service failure is not resolved. In short, when a customer experiences incomplete service recovery, the benefits associated with social presence cues do not outweigh the responsiveness perception. Therefore, embedding social presence cues is not a 'cure-all solution' for increasing customer satisfaction.

In summary, social presence cues are a proxy for emotional exchanges that typically take place in face-to-face interactions, and such cues create a virtual mood. The findings of this study suggest that social presence cues (1) create positive experiences in virtual interactions and (2) improve customers' overall attitudes toward the firm, particularly when service failure is not resolved. Understanding the importance of social presence cues will allow firms to strategically plan mobile communications that convey empathy and human warmth.

Managerial implications

One important implication for marketers is the identification of new technologies, such as texting via social media platforms, to provide a prompt means of resolving service failure. Global technologies give customers more power and influence, and customers can easily take action using their mobile device when they are dissatisfied with services. To reduce the spread of negative word of mouth, firms should offer a convenient means for customers to directly address their complaints with the firm. Identifying effective intervention strategies when service failure occurs is an important step in maintaining high satisfaction levels among customers. For example, a dissatisfied customer can negatively affect a retailer's image by spreading harmful and destructive content via the web. Negative word of mouth can be detrimental for firms, especially for retailers that rely heavily on positive online reviews. When negative claims go viral, marketers have little control over the content, how it is perceived, and whom it reaches. Therefore, addressing customers' concerns and complaints in a timely and efficient manner is a necessity for marketers operating in a world in which consumers are virtually connected. Responding expediently to customers' complaints, concerns, or queries with the inclusion of social presence cues is a valuable strategy when the target demographic is heavy smartphone users.

In general, marketing communications accomplish two objectives: delivering content and establishing relationships with the other party. When messages are automated and inauthentic, a negative mood may arise among receivers. As the focus group participants conveyed, the lack of social cues in text messages is similar to 'bad customer service in retail departments'. Adding social presence cues to text messages will provide a positive virtual environment in which online interactivity influences offline attitudes and behaviors. More important, messages that include social presence cues are particularly important when problems are not resolved in a timely manner.

The study's findings show no negative effects of social presence cues; thus, firms can apply social presence cues in all types of communications. Social presence cues had the greatest impact on two-way communication perceptions and satisfaction (see the effect sizes in Table 2). Although social presence cues can be used to improve the perceived value of mobile technology communications, these cues alone will not increase customers' attitudes and repurchase intentions. Social presence cues can enhance customer satisfaction; however, the primary goal for marketers must be successful and effective service recovery. It is our hope that marketers will apply the findings of this study by including social presence cues in all virtual communications as a means for improving customer satisfaction and overall attitude toward the firm.

Theoretical implications

The theoretical contributions of the study are threefold. First, we apply social presence theory in the context of real-time mobile texting via Facebook. Previous studies have explored the influence of social presence cues on trust (Gefen & Straub, 2003, 2004), system/media acceptance (Karahanna & Straub, 1999), and patronage behaviors (Hassanein & Head, 2005) on commercial websites. With significant empirical evidence, this study provides insights into the importance of social presence theory in the context of m-commerce.

Second, whereas prior research on social presence cues has demonstrated the effect of nonverbal cues (e.g. avatar and emoticon) or textual cues (e.g. information on the presence of other customers) on interactivity perceptions and communication effectiveness, the current study focuses on the effect of verbal cues in mobile texting communications. Thus, the findings of this study contribute to the literature on social presence in computer-mediated communications by examining social cues in mobile text messages among heavy smartphone users.

Third, our study sheds light on ways to manipulate interactive communications. Various interactivity features have been tested and discussed in prior studies (Ha & James, 1998; Macias, 2003). Some features are strong predictors of two-way communication perceptions (e.g. feedback mechanism), control perceptions (e.g. search map), or responsiveness perceptions (e.g. speed). Our study demonstrates that verbal social presence cues stimulate interactivity perceptions. In particular, social presence cues are strong determinants of two-way communication perceptions. Therefore, adding social presence cues to text messages is an effective and efficient way to manipulate interactive messages.

Directions for further research

Further research can build on the study's findings in several areas. First, we only consider one new media channel: real-time mobile texting. Additional research could examine communications within other emerging channels, such as online communities, Facebook, Twitter, or blogs. Second, we focused on verbal cues in texting situations. However, what if the image or picture of a corresponding customer representative is included in the text? Various types of social cues might influence interactivity perceptions differently. Although the results of this study clearly indicate that verbal cues have the strongest impact on two-way communication, visual cues (e.g. image and emoticon) might stimulate other interactivity perceptions (e.g. responsiveness). Cui et al. (2010) test two types of social cues (i.e. cognitive cue and affective cue) on interactivity perceptions but do not examine and compare effect sizes. Future studies could test various social cues and the relative influence on all three interactivity perceptions.

Third, in terms of manipulating the level of social presence, we only consider two levels: with and without social presence cues. Future studies might examine the impact of various levels of social presence cues. For example, consumers might feel irritated or uneasy under extremely high levels of social presence. It would be fruitful to determine the optimal level of social presence in various computer-mediated communication situations. Fourth, we used a student sample. Although younger generations are leading the industry in adopting and using mobile technologies, further research should consider using a representative sample with diverse demographics.

Mobile technologies provide an important means for marketers in establishing interactive communications with their customers. A first step toward stimulating mobile communications is to understand customers' value perceptions. While customers value time and convenience, they also value human warmth.

Disclosure statement

No potential conflict of interest was reported by the authors.

References

Balasubramanian, S., Perterson, R. A., & Jarvenpaa, S. L. (2002). Exploring the implications of m-commerce for markets and marketing. *Journal of the Academy of Marketing Science*, *30*(4), 348–361.

Bird, C. (2013). *In-car electronics: Entertainment and navigation*. Mintel Report. Retrieved from http://store.mintel.com/in-car-electronics-entertainment-and-navigation-us-august-2013

Cambra-Fierro, J., Berbel-Pineda, J. M., Ruiz-Benítez, R., & Vázquez-Carrasco, R. (2013). Analysis of the moderating role of the gender variable in service recovery processes. *Journal of Retailing and Consumer Services*, *20*(4), 408–418.

Chang, H.-S., & Hsiao, H.-L. (2008). Examining the casual relationship among service recovery, perceived justice, perceived risk, and customer value in the hotel industry. *Service Industries Journal*, *28*(4), 513–528.

Chan-Olmsted, S., Rim, H., & Zerba, A. (2013). Mobile news adoption among young adults: Examining the roles of perceptions, news consumption, and media usage. *Journalism & Mass Communication Quarterly*, *90*(1), 126–147.

Chelminski, P., & Coulter, R. A. (2011). An examination of consumer advocacy and complaining behavior in the context of service failure. *Journal of Services Marketing*, *25*(5), 361–370.

Chen, K.-Y., & Chang, M.-L. (2013). User acceptance of 'near field communication' mobile phone service: An investigation based on the 'unified theory of acceptance and use of technology' model. *Service Industries Journal*, *33*(6), 609–623.

Chhateja, J., & Jain, V. (2014). Understanding generation Y and their perspective on proximity and permission based SMS marketing. *Romanian Journal of Marketing*, (4), 2–10. Retrieved from http://www.readperiodicals.com/201410/3608581791.html

Cowles, D., & Crosby, L. A. (1990). Consumer acceptance of interactive media in service marketing encounters. *Service Industries Journal*, *10*(3), 521–540.

Coyle, J. R., & Thorson, E. (2001). The effects of progressive levels of interactivity and vividness in web marketing sites. *Journal of Advertising*, *30*(3), 65–77.

Cui, N., Wang, T., & Xu, S. (2010). The influence of social presence on consumers' perceptions of the interactivity of web sites. *Journal of Interactive Advertising*, *11*(1), 36–49.

Dabholkar, P. A., & Spaid, B. I. (2012). Service failure and recovery in using technology-based self-service: Effects on user attributions and satisfaction. *Service Industries Journal*, *32*(9), 1415–1432.

Daft, R. L., & Lengel, R. H. (1986). Organizational information requirements, media richness, and structural design. *Management Science*, *32*(5), 554–571.

DeWitt, T., Nguyen, D. T., & Marshall, R. (2008). Exploring customer loyalty following service recovery. *Journal of Service Research*, *10*(3), 269–281.

Fornell, C., Johnson, M. D., Anderson, E. W., Cha, J., & Bryant, B. E. (1996). The American customer satisfaction index: Nature, purpose, and findings. *Journal of Marketing*, *60*(4), 7–18.

Fulgoni, G., & Lipsman, A. (2014). Digital game changers: How social media will help usher in the era of mobile and multi-platform campaign-effectiveness measurement. *Journal of Advertising Research*, *54*(1), 11–16.

Gefen, D., & Straub, D. (2003). Managing user trust in B2C e-services. *e-Service Journal*, *2*(2), 7–24.

Gefen, D., & Straub, D. (2004). Consumer trust in B2C e-commerce and the importance of social presence: Experiments in e-products and e-services. *Omega*, *32*, 407–424.

Gorn, G. J., Chattopadhyay, A., Sengupta, J., & Tripathi, S. (2002). Waiting for the web: How screen color affects time perception. *Journal of Marketing Research*, *41*(2), 215–225.

Ha, L., & James, L. (1998). Interactivity reexamined: A baseline analysis of early business web sites. *Journal of Broadcasting and Electronic Media*, *42*(4), 457–474.

Hair, J. F., Black, W. C., Babin, B. J., & Anderson, R. E. (2009). *Multivariate data analysis*. Upper Saddle River, NJ: Prentice Hall.

Harland, B. (2015). *Mobile phone*. Mintel Reports. Retrieved from http://store.mintel.com/mobile-phones-us-february-2015

Harrison-Walker, L J. (2012). The role of cause and affect in service failure. *Journal of Services Marketing*, *26*(2), 115–123.

Hassanein, K., & Head, M. (2005). The impact of infusing social presence in the web interface: An investigation across product type. *International Journal of Electronic Commerce*, *10*(2), 31–55.

Holzwarth, M., Janiszewski, C., & Neumann, M. M. (2006). The influence of avatars on online consumer shopping behavior. *Journal of Marketing*, *70*(4), 19–36.

Homburg, C., & Fürst, A. (2005). How organizational complaint handling drives customer loyalty: An analysis of the mechanistic and the organic approach. *Journal of Marketing*, *69*(3), 95–114.

Jiangang, D., Fan, X., & Feng, T. (2011). Multiple emotional contagions in service encounters. *Journal of the Academy of Marketing Science*, *39*(3), 449–466.

Johnson, G. J., Bruner II, G. C., & Kumar, A. (2006). Interactivity and its facets revisited. *Journal of Advertising*, *35*(4), 35–52.

Karahanna, E., & Straub, D. W. (1999). The psychological origins of perceived usefulness and ease-of-use. *Information and Management*, *35*(4), 237–250.

Kiousis, S. (2002). Interactivity: A concept explication. *New Media & Society*, *4*(3), 355–383.

Kleijnen, M., De Ruyter, K., & Wetzels, M. M. (2007). An assessment of value creation in mobile service delivery and the moderating role of time consciousness. *Journal of Retailing*, *83*(1), 33–46.

Liu, Y. (2003). Developing a scale to measure the interactivity of websites. *Journal of Advertising Research*, *43*(3), 207–216.

Macias, W. (2003). A beginning look at the effects of interactivity, product involvement, and web experience on comprehension: Brand web sites as interactive advertising. *Journal of Current Issues and Research in Advertising*, *25*(2), 31–44.

McGee, M. J. (2014). Is texting ruining intimacy? Exploring perceptions among sexuality students in higher education. *American Journal of Sexuality Education*, *9*(4), 404–427.

McMillan, S. J., & Hwang, J. (2002). Measures of perceived interactivity: An exploration of the role of direction of communication, user control, and time in shaping perceptions of interactivity. *Journal of Advertising*, *31*(3), 29–42.

Moore, R., & Moore, M. (2004). Customer inquiries and complaints: The impact of firm response time to email communications. *Marketing Management Journal*, *14*(2), 1–12.

Olmsted, N. M., & Terry, C. P. (2014). Who's texting in class? A look at behavioral and psychological predictors. *Psi Chi Journal of Psychological Research*, *19*(4), 183–190.

Pan, S. L., & Lee, J. (2003). Using e-CRM for a unified view of the customer. *Communications of the ACM*, *46*(4), 95–99.

Preacher, K. J., & Hayes, A. F. (2004). SPSS and SAS procedures for estimating indirect effects in simple mediation models. *Behavior Research Methods, Instruments, and Computers*, *36*(4), 717–731.

Pynta, P., Seixas, S. A. S., Nield, G. E., Hier, J., Millward, E., & Silberstein, R. B. (2014). The power of social television: Can social media build viewer engagement? A new approach to brain imaging of viewer immersion. *Journal of Advertising Research*, *54*(1), 71–80.

Roschk, H., & Kaiser, S. (2013). The nature of an apology: An experimental study on how to apologize after a service failure. *Marketing Letters*, *24*(3), 293–309.

Sago, B. (2010). The influence of social media message sources on millennial generation consumers. *International Journal of Integrated Marketing Communications*, *2*(2), 7–18.

Shankar, V., Inman, J. J., Mantrala, M., Kelley, E., & Rizley, R. (2011). Innovations in shopper marketing: Current insights and future research issues. *Journal of Retailing*, *87*(1), S29–S42.

Short, J. A., Williams, E., & Christie, B. (1976). *The social psychology of telecommunications.* New York, NY: John Wiley & Sons.

Song, J., & Zinkhan, G. M. (2008). Determinants of perceived web site interactivity. *Journal of Marketing, 72*(2), 99–113.

Spiggle, S. (1994). Analysis and interpretation of qualitative data in consumer research. *Journal of Consumer Research, 21*(3), 491–503.

Steuer, J. (1992). Defining virtual reality: Dimensions determining telepresence. *Journal of Communication, 42*(4), 73–93.

Strauss, J., & Hill, D. J. (2001). Consumer complaints by e-mail: An exploratory investigation of corporate response and customer reaction. *Journal of Interactive Marketing, 15*(1), 63–73.

Thomas, D. (2014). The mobile consumer revealed. *Marketing Magazine, 119*(5), 38–39.

Tsai, C.-T., & Su, C.-S. (2009). Service failures and recovery strategies of chain restaurants in Taiwan. *Service Industries Journal, 29*(12), 1779–1796.

Williams, E. (1977). Experimental comparisons of face-to-face and mediated communication: A review. *Psychological Bulletin, 84*(5), 963–976.

Wu, G. (2005). The mediating role of perceived interactivity in the effect of actual interactivity on attitude toward the website. *Journal of Interactive Advertising, 5*(2), 29–39.

Yen, H. R. (2005). An attribute-based model of quality satisfaction for internet self-service technology. *Service Industries Journal, 25*(5), 641–659.

Zeithaml, V. A., Berry, L. L., & Parasuraman, A. (1996). The behavioral consequences of service quality. *Journal of Marketing, 60*(2), 31–46.

Zhao, X., Lynch, J. G., & Chen, Q. (2010). Reconsidering Baron and Kenny: Myths and truths about mediation analysis. *Journal of Consumer Research, 37*(2), 197–206.

Customer e-complaining behaviours using social media

M. S. Balaji[a], Subhash Jha[b] and Marla B. Royne[c]

[a]Taylor's Business School, Taylor's University, Selangor, Malaysia; [b]Department of Marketing, Indian Institute of Management, Udaipur, India; [c]Department of Marketing and Supply Chain Management, Fogelman College of Business and Economics, The University of Memphis, Memphis, TN, USA

This paper develops a conceptual framework about customer complaining behaviours (CCB), using social media. Specifically, this research expands the current understanding of CCB by examining the differential impact of unfairness, firm response, retaliation, locus attribution, stability attribution, and personal identity on public complaining and private complaining using social media, and their subsequent impact on post-complaining satisfaction (PCS) and loyalty. Public complaining refers to customer complaints directed to a service provider, while private complaining refers to service failure complaints directed towards other customers. A structural equation model shows that high levels of unfairness, firm response, locus, and personal identity have a strong influence on public complaining, while desire for retaliation is a significant factor influencing private complaining. The findings contribute to practice by providing useful and pertinent information for developing suitable web care interventions to effectively deal with public complaining and private complaining through social media platforms.

Introduction

Service failures are inevitable due to the unique characteristics and complexities involved in service delivery. When such failures occur, customers engage in complaining behaviours to resolve the problem (Hong & Lee, 2005). Although customers may employ a number of different channels to communicate their complaint, such as face-to-face contact, the telephone, postal mail, the recent emergence of social media has empowered many customers to complain online in seeking redress or venting frustration (Tripp & Grégoire, 2011). Before the advent of social media, a majority of dissatisfied customers failed to complain, as the cost of complaining was perceived to far exceed the benefits associated with service recovery (Sharma, Marshall, Alan Reday, & Na, 2010). This has dramatically changed with the emergence of social media. Customers can now use social media platforms to complain more directly, conveniently, and effectively than ever before to the service provider. Moreover, if customers do not receive a quick response, the complaint can be moved to a public domain, causing major problems for the service provider (Grégoire, Salle, & Tripp, 2014).

Thus, social media has changed the dynamics of communication between service providers and customers, transforming the way customers communicate their complaints.

Although social media has attracted considerable attention from researchers in recent years, empirical work investigating customer complaining behaviours (CCB) using social media remains scarce. For example, Clark (2013) noted that there is very little research regarding the role of social media platforms in complaints against service providers. In addressing this gap, this paper examines the antecedents and consequences of CCB using social media. Specifically, this study has used Day and Landon's (1977) taxonomy of complaining behaviours to investigate factors influencing public complaining and private complaining behaviours via social media. The term public complaining refers to customer complaints of a service failure directed to the service provider, while the term private complaining includes customer complaints of a service failure directed at other customers (Day & Landon, 1977; Harrison-Walker, 2001). For instance, on the Facebook social media platform, public complaining involves customers posting their complaints on the service providers' Facebook page. In contrast, when customers complain about a service failure to their friends and acquaintances through sharing on their personal Facebook page, this may be referred to as private complaining. Under these definitions, public complaints are visible for the service provider to see, while private complaints largely remain undetected by the service provider. In attempting to gain an understanding of the factors that determine customers' public and private complaining behaviours using social media, this study considers the six following possible antecedents: perceived unfairness, likelihood of firm response, desire for retaliation, stability attribution, locus attribution, as well as personal identity based on the underpinnings of the justice theory, the attribution theory, and the self-categorization theory (SCT). This study also explores the outcomes of complaining by the customer using social media on post-complaining satisfaction (PCS) and loyalty.

This research study makes three important contributions to the service failure literature. First, the study extends the understanding of CCB using social media channels by applying three psychological theories to identify relevant factors that motivate customers to complain about a service failure using social media. Second, this research has delineated between public and private complaining using social media in understanding CCB. Finally, this study examines the outcomes of CCB via social media on PCS and loyalty. Findings will assist service managers to enhance their understanding of the determinants of CCB using social media. The findings will also aid in the development of appropriate web care intervention strategies to effectively deal with customer complaints made through social media.

Theoretical background

This study draws on the key aspects of the justice theory, the attribution theory, and the SCT to develop a conceptual model of CCB using social media following a service failure. A key objective of this paper is the integration of these theories into a cohesive framework describing customers' public complaining and private complaining on social media.

The justice theory, or the fairness perception theory, relates to an individual's perception of justice or fairness in an encounter (Greenberg, 1987). This theory posits that when customers perceive injustice or imbalance, they feel dissatisfied and take actions to remedy the problem. In a service failure encounter, when customers perceive loss, they engage in complaining behaviours to seek redress or reprisal (Tax, Brown, & Chandrashekaran, 1998). Thus, the justice theory provides a useful theoretical framework to understand the role of perceived fairness during service failures in CCB.

While the justice theory links fairness to complaining behaviours, the attribution theory complements it by examining the customer's beliefs about the causes of a service failure. The attribution theory states that customers are rational information processors whose behaviours are influenced by causal inferences for a service failure (Heider, 1958). These causal inferences for a service failure may influence the CCB. As accountability of a service failure plays a key role in fairness perception (McColl-Kennedy & Sparks, 2003), the attribution theory complements the justice theory in understanding CCB using social media.

SCT holds that customers define themselves as both individuals and as member of a group (Turner, 1985), and this process allows customers to express their self-concept or personal identity. According to this theory, customers are motivated to engage in activities that promote their identity. Because social media provides people a way to define their self-concept (Kaplan & Haenlein, 2010), SCT has been used in this study to help understand the role of personal identity in CCB using social media.

While the justice theory and the attribution theory focus on service failure evaluation, SCT relates to the role of customers' self-concept in their behavioural engagements. That is, while justice and attribution research examines the role of situational factors such as perceived unfairness and attribution of blame, SCT looks beyond such factors to examine the customers' self-concept in complaining behaviours. As the three theories are complementary, they may be used together to examine the CCB using social media following a service failure. Moreover, as an individual's identity in a group is shaped by justice perceptions (Tyler & Blader, 2003) and causal explanations for events (Weary & Arkin, 1981), SCT complements the justice theory and the attribution theory in extending our understanding of CCB using social media.

Justice theory

According to the justice theory, individuals strive for justice in an exchange situation. By comparing the ratio of the outcomes (what an individual receives) to the inputs (what an individual pays) relative to others' output and input ratio, individuals determine the perceived justice in an exchange. When an imbalance forms, the exchange parties will try to address and rectify the situation and bring it to a state of equilibrium (Tax et al., 1998). Scholars have noted that customers tend to experience loss and perceive inequity in a service failure encounter (McCollough, Berry, & Yadav, 2000). In such cases, fair or just recovery restores equity and shapes customer responses to a service failure. Here, the level of satisfaction depends on the extent to which a customer perceives being treated fairly in a service encounter. Failure to provide a sense of justice may lead to perceptions of contract violation, further resulting in negative responses such as emotional venting, rage, switching behaviour, negative word-of-mouth, and boycotting the service provider (Gelbrich, 2010; Weun, Beatty, & Jones, 2004).

Based on the justice theory, Blodgett, Hill, and Tax (1997) examined the role of perceived justice on post-complaint behaviours in a retail store environment. The authors found that when customers perceive injustice, they become angry. This impacted both their repatronage and negative word-of-mouth intentions. Similarly, Park, Kim, and O'Neill (2014) used the justice theory to understand CCB and service recovery across different cultures. While significant differences were observed between complaint behaviours in individualistic and collectivistic cultures, the authors suggested that dissatisfaction with a service failure ultimately determines whether consumers resort to complaining directly to the service provider or to complaining to acquaintances. More recently, Wu and Huang (2015) have demonstrated that perceived justice determines complaint intentions after online service failures. In summary, research demonstrates that the justice theory provides a useful framework to understand CCB using social media following a service failure.

Attribution theory

The attribution theory deals with how individuals understand an event, as well as how this understanding influences their decision making. According to Heider (1958), individuals often engage in *post hoc* causal searches to understand behaviours (their own and others) or events. Based on the information search, along with their internal beliefs and motives, individuals make causal inferences for the event or behaviour, which determines their subsequent responses. Eberly, Holley, Johnson, and Mitchell (2011) argued that individuals often make multiple attributions to a particular event or behaviour. When service failures occur, customers evaluate the failure in terms of its likelihood to recur (stability), who is to blame (locus), and if it could be avoided (controllability). If a customer perceives the failure as his or her fault, less likely to recur, or unavoidable, then he or she is more likely to be satisfied and remain loyal to the service provider (Hess, Ganesan, & Klein, 2003). The present study considers both stability and locus attributions, as they are most salient in understanding the CCB.

The justice theory and the attribution theory have been used in tandem in the service failure literature to understand customer response behaviours. For example, Blodgett, Wakefield, and Barnes (1995) developed a dynamic model of CCB. Based on both the justice theory and the attribution theory, the authors argued that stability attribution, controllability attribution, and perceived justice dimensions influence redress and negative word-of-mouth. Similarly, Tojib and Khajehzadeh (2014) determined that attribution (firm attribution versus self-attribution) and perceived severity of the failure (extent of loss) have an influence on customers' direct complaint intentions. The authors further proposed that attribution and justice perceptions play a key role in understanding why customers directly complain to the service provider. Consequently, both the justice and the attribution theory have been used in the current study to understand CCB using social media.

Self-categorization theory

SCT posits that individuals define their self-concept through the placement of self and others within a social environment (Turner, 1985). According to this theory, self-concept operates at three levels. Human identity operates at the super-ordinate level, while social identity operates at the intermediate level, and the personal identity operates at the subordinate level of the particular individual (Hornsey, 2008). Social identity is described in terms of the group membership and social standing of an individual in the identified group. Personal identity, on the other hand, is self-conception based on idiosyncratic characteristics or one's interpersonal relationship with others. Extant literature suggests that social identity becomes more prominent when an individual is characterized as belonging to a group and personal identity overrides social identity in situations of an intragroup context (Turner, 1985).

Previous research suggests that self-identity serves as a source of information and may have a pervasive influence on an individual's intentions and behaviours (Settles, Jellison, & Pratt-Hyatt, 2009). Drawing on SCT, the current study proposes that in a social group (e.g. Facebook), customers construct context-specific identity or prescriptions that influence

70

their attitudes and behaviours, including complaining behaviours. Because self-categorization depends on both accessibility (past experience, present expectations, and current motives) and fit (extent to which an individual perceives fit with the group) (Hogg & Terry, 2000), it may be argued that a customer's personal identity may be important in communicating their complaints using social media. The relative accessibility of the service failure encounter, along with the need to enhance or maintain their self-concept and affiliation with the group, may prompt customers to engage in complaining behaviours using social media. Consequently, the present study considers personal identity in understanding CCB using social media.

Only a few works in the literature have examined the role of SCT in CCB. For example, Ward and Ostrom (2006) proposed that customer identity might influence the service failure responses. Wan (2013) argued that face concerns reflects self-concept of an individual and impacts customer complaint intentions, switching, and negative word-of-mouth behaviours. These studies thus suggest that self-concept or identity may determine customer complaint intentions. In a social psychology context, Van Stekelenburg and Klandermans (2013) showed that perceived justice and self-categorization influence protest behaviours. Similarly, Costarelli (2009) demonstrated that group identification and causal attributions jointly impact an individual's experienced affect. As justice and attribution perceptions are motivated by instrumental concerns (fair outcomes) and relational outcomes (identity in a group), SCT complements the justice theory and the attribution theory in understanding CCB using social media.

Research model and hypotheses

Based on the three theoretical domains, this study has developed a conceptual model (Figure 1) proposing that complaining behaviours using social media are determined by situational and personal factors such as unfairness, locus attribution and stability attribution, desire for revenge, likelihood of firm response, and personal identity. More specifically, this study examines the differential impact of these factors on customers' public and private

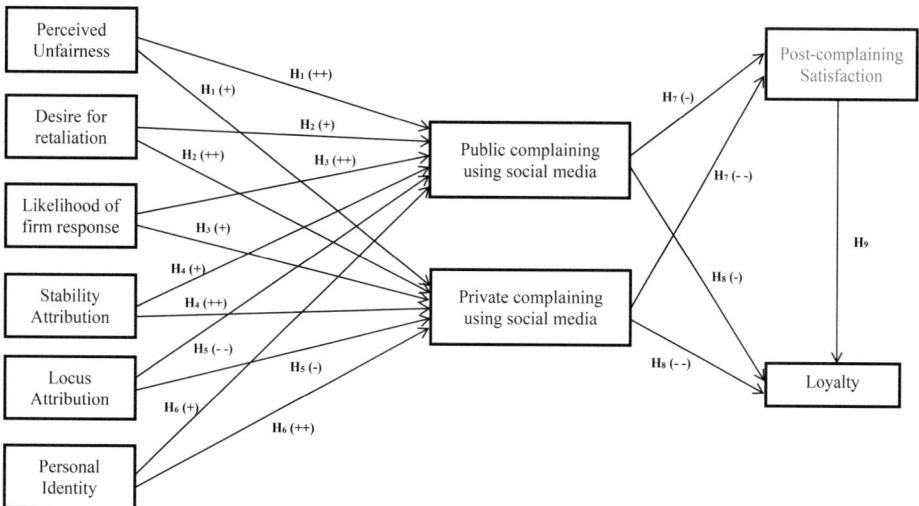

Figure 1. Conceptual framework of the customer complaining behaviours using social media. Note: ++ sign indicates greater positive effect than +. − − sign indicates greater negative effect than −.

complaining behaviours via social media. The relationships between CCB, PCS, and loyalty have also been explored.

Perceived unfairness

The justice theory holds that the perceived injustice or unfairness underlies a customer's response to a service failure. As noted, perceived unfairness relates to the extent to which the customer feels that he/she has been treated unfairly or unjustly in a service encounter. Many researchers have examined the effect of unfairness (magnitude of the failure or severity of the failure) on CCB. Research shows that complaining to the service provider (public) or to friends and others (private) increases when perceived unfairness is greater (Balaji & Sarkar, 2013; Casado-Diaz & Nicolau-Gonzalbez, 2009). Swanson, Frankel, Sagan, and Johansen (2011) showed that during high unfairness encounters, customers perceive greater inequity and are more likely to seek redress by complaining to the service provider. Public complaining provides an opportunity for the service provider to rectify the error and potentially create a more satisfied customer. From an economic perspective, as customers expect fair products or services for the money they have spent, public complaining may restore the imbalance. On the contrary, when customers engage in private complaining, the service provider is unaware of the service failure and is typically unable to resolve or rectify the problem (Bodey & Grace, 2007). Therefore, customers are more likely to complain directly to a service provider than to inform friends or acquaintances when they experience unfairness. Hence, the following is hypothesized:

> *H1:* Perceived unfairness has a more positive effect on public complaining using social media than private complaining using social media.

Desire for retaliation

Retaliation refers to customer actions and behaviours made with intent to penalize or cause inconvenience to the service provider for losses suffered. Retaliatory behaviour may be explained by the justice theory because a service failure is perceived as a violation of fairness and trust norms, creating a sense of betrayal among customers who have strong relationships with their service provider (Wang & Huff, 2007). Researches indicated that perceived betrayal in a service failure encounter motivates customers to engage in retaliation behaviours. Grégoire and Fisher (2006) found a direct effect of desire for retaliation on negative word-of-mouth, third-party complaining, and reduction in patronage. Particularly, retaliation had a greater effect on negative word-of-mouth than patronage reduction. By engaging in negative word-of-mouth to friends and others, customers malign the service provider's reputation and report their negative experience to a wider audience (Gelbrich, 2010; Sparks & Browning, 2010). The underlying goal is to harm the service provider by encouraging others not to patronize it. Hence, the following hypothesis is proposed:

> *H2*: Desire for retaliation has a more positive effect on private complaining using social media than public complaining using social media.

Likelihood of firm response

Likelihood of firm response is the customer's perception of the service provider's willingness to remedy the service failure (e.g. compensation, apology) (Orsingher, Valentini, & de Angelis, 2010). Previous research has shown that likelihood of firm response is positively related to CCB. It has also argued that likelihood of firm response is a trust-creating behaviour which affects the perception of injustice and customers' behavioural responses in a service failure (Chiu, Chou, & Chiu, 2013). Thus, based on the justice theory, we propose that when customers believe the service providers are responsive in handling the complaints, they are likely to complain to the providers. When a service provider is perceived as unresponsive, this restricts customers' complaint behaviours. In this case, customers are more likely to inform friends or others. Hence, the following is proposed:

> *H3*: Perceived likelihood of firm response has a more positive effect on public complaining using social media than private complaining using social media.

Attribution of stability

Stability refers to the extent to which customers perceive the reasons for a service failure as enduring or permanent (Hess et al., 2003). Service failure literature suggests that stability attribution affects several key affective and behavioural outcomes. Using the attribution theory, Kaltcheva, Winsor, and Parasuraman (2013) found that customer repatronage intentions depend on the perceived stability of the service failure. When customers perceive a stable cause for a service failure, they expect to suffer a similar outcome in the future. This increases their anger and leads to higher dissatisfaction and negative behavioural outcomes (Swanson & Hsu, 2011). In contrast, ascribing a service failure to an unstable cause implies that a future occurrence of such events is uncertain. Therefore, customers are more likely to voice their complaint to the service provider. Blodgett et al. (1995) proposed that customers who believed the reasons for the service failure as stable were more likely to engage in negative word-of-mouth than to repatronize the service provider. Hence, it has been posited that perceived stability leads to unfavourable evaluations because customers tend to believe that similar failures may reoccur in the future, resulting in their engaging in negative word-of-mouth with friends and acquaintances. Therefore, the following is proposed:

> *H4*: Stability attribution has a more positive effect on private complaining using social media than public complaining using social media.

Attribution of locus

Attribution of locus refers to the extent to which customers perceive the reason or reasons for failures as related to self, the service provider, or external factors (Swanson & Hsu, 2011). While prior studies have excluded locus attribution in examining complaining behaviour, it is relevant here because it provides better understanding of the CCB using social media. The self-serving bias in the attribution theory postulates that depending on their level of participation in the service exchange, customers may attribute the blame to themselves. In this case, customers feel they are at fault, expect little or no recovery, and are less likely to engage in confrontive behaviours. Mattila and Ro (2008) found that self-blame reduces intentions to generate negative word-of-mouth, while attribution of locus fails to influence complaints to the service provider. Harris, Mohr, and Bernhardt (2006) found that customers blamed themselves for service failures in the online shopping environment and expected lower levels of service recovery. They feel guilty and accept the service failure as reparation, inhibiting the complaining behaviour and word-of-mouth communication. Therefore, it is hypothesized:

> *H5*: Higher levels of locus attribution (self-blame) have a more negative effect on public complaining using social media than private complaining using social media.

Personal identity

Personal identity refers to an individual's sense of self. It includes personal characteristics such as attributes, abilities, traits, and interests (Turner, 1985). Although personal identity influences the evaluation of, and behaviour towards the service provider, little research has assessed its role in the service failure context. Haslam, Branscombe, and Bachmann (2003) examined the role of organizational identification in their evaluation of a service failure, finding that customers who strongly identify with the service provider are more likely to forgive the service failure and engage in repeat purchases than those who do not identify with the organization.

Social media allow individuals to signal meaningful cues about themselves by displaying their personal self to others (Kaplan & Haenlein, 2010). Because personal identity concerns a customer's idiosyncratic qualities and abilities, high personal identity customers are more likely to engage in private complaining using social media to express their dissatisfaction. Support for this argument is found in the self-consciousness literature. Marquis and Filiatrault (2002) found that self-consciousness or disposition to focus on self affects CCB. Specifically, in a service encounter, high self-conscious customers favour negative word-of-mouth as a preferred complaining response over directly reporting to the service provider. The authors argued that the greater concern for impression management among the high self-consciousness customers makes them prefer private actions; they are reluctant to react publicly no matter the cause of the service failure. Thus, high personal identity customers are more likely to complain through private than public channels. Hence, the following hypothesis is offered:

> *H6*: Higher levels of personal identity have a more positive effect on private complaining using social media than public complaining using social media.

CCB, PCS, and loyalty

PCS refers to the satisfaction customers derive from the act of complaining (Bearden & Oliver, 1985). Research has shown that complaining behaviour is a key precursor for the service provider to remedy the problem and recover customer satisfaction. Customers who complain via public channels may expect a remedy or resolution of the service failure. In contrast, private complaining denies the service provider the opportunity to resolve a service failure (Swanson et al., 2011). Bearden and Oliver (1985) found an inverse relationship between private complaining and satisfaction, meaning that greater private complaining resulted in lower customer satisfaction, and public complaining was positively related to satisfaction with firm response. Existing research has further shown that effective complaint handling is related to satisfaction and future repurchase intentions (Tax et al., 1998). Hence, the following replication hypothesis in the social media context is offered:

> *H7*: Higher levels of private complaining via social media have a more negative effect on post-complaining satisfaction than public complaining using social media.

Prior research has indicated that CCB negatively affects the loyalty towards service provider. Holloway and Beatty (2003) offer indirect support for this relationship; they found that only 25.6% of the respondents who complained to the service provider planned to return to that company while 74.4% indicated they would not repurchase from that company again. Rothenberger, Grewal, and Iyer (2008) reported that the simple act of complaining is negatively related to the loyalty dimensions of recommendation and reuse. The authors argued that loyalty depends on fair resolution of the service problem. An ineffective solution to a customer complaint leads to exit or negative loyalty behaviour.

Customers engage in public complaining because they believe such complaints will be handled appropriately. This motivates their desire to correct the problem, rather than dissolve the relationship with the service provider. In contrast, private complaining reflects the decline or absence of a customer–service-provider relationship (Blodgett et al., 1995). Hence, the following replication hypothesis in social media context is proposed:

> *H8*: Higher levels of private complaining via social media have a more negative effect on loyalty than public complaining using social media.

Prior research has empirically demonstrated that satisfaction is strongly related to loyalty (Swanson & Hsu, 2011). Specifically, research shows that loyalty, word-of-mouth, and trust are the results of PCS and that satisfaction is positively related to loyalty. Extending this to the current context, the following replication hypothesis is offered in the social media context:

> *H9*: Post-complaining satisfaction is positively related to loyalty.

Method

Sample

Data were collected using a structured questionnaire. Facebook was chosen as study context for examining the complaining behaviours using social media. A recent report revealed that Facebook has about 1.2 billion monthly active users. More than half of them have shared their product-related experiences with others on the social networking platforms (Logan, 2014). While previous studies have conceptualized social media platforms as a complaint channel, the current research empirically examines the antecedents and consequences of CCB using social media.

A sample of 176 undergraduate and graduate business students at a large private university was obtained using a convenience sampling method. A student sample is appropriate for the following reasons: (1) demographics for social media tend to be young; (2) students are heavy consumers of social media platforms; and (3) the use of college students is consistent with previous research on social media (Alexandrov, Lilly, & Babakus, 2013; Lim & Palacios-Marques, 2011). Moreover, Facebook is a popular social networking platform among young adults for sharing their brand-related experiences in the form of posts, pictures, comments, and ratings (Clark, 2013). Finally, there is precedent for use of a student sample in service failure research (Kaltcheva et al., 2013). Screening questions were used to ensure that only students with a hotel experience during the past six months and who were current active users of Facebook completed the survey. Students participated voluntarily in the study and were assured of confidentiality. A comparison of sample demographic characteristics to the country internet users indicated that the respondents were representative of the part of the online population who are extensive users of social media.

Procedure

All participants read a scenario of a service failure in a hotel setting (see Appendix) and then completed the questionnaire in a self-administered manner in a classroom. A pre-test was performed to develop the service failure scenario. Every effort was made to ensure that the service failure scenario was realistic and comprehensible to a student sample. Participants were told that the authors were interested in their reactions to a hotel service failure experience. After reading the scenario, participants indicated the realism and comprehensibility of the scenario on two nine-point scales, in which a higher score indicates a more favourable response. The realism ($Ms > 7.13$) and comprehensibility ($Ms > 6.16$) ratings were high. All respondents were members of at least one of several different social networking websites, including Facebook, Twitter, LinkedIn, Hangout, My Space, and Tagged. Each respondent had a Facebook account and almost all used it at least once in a day (94.9%). All respondents were between age 18 and 27, and nearly half (45%) were female.

Measurement

Study constructs were operationalized via multi-item scales. Perceived unfairness was measured with a three-item scale taken from Oliver and Swan (1989). Desire for retaliation was operationalized via five items borrowed from Grégoire and Fisher (2006). Likelihood of firm response was measured using three items adopted from Kim, Kim, Im, and Shin (2003). Three items from Weiner (1985) were used to assess attribution stability. Locus attribution was captured via two items borrowed from Garnefski and Kraaij (2007). Glynn's (2008) three-item scale was used to measure personal identity. Public complaining intentions and private complaining intentions were each measured with three items adapted from Wan (2013). PCS was measured using three items from McCollough et al. (2000), while loyalty was assessed with a three-item scale from Zboja and Voorhees (2006). All constructs were measured with seven-point scales, such that higher scores consistently indicated higher levels of each construct.

Analysis and results

Measurement results

Because of the single source of data, the potential for common method bias was investigated using Harman's one-factor test (Podsakoff, MacKenzie, Lee, & Podsakoff, 2003). First, a ten-factor measurement model ($\chi^2 = 640.67$, $df = 417$, $\chi^2/df = 1.53$) was tested followed by a single factor measurement model ($\chi^2 = 3044.10$, $df = 434$, $\chi^2/df = 7.01$). The results of the χ^2 difference test ($\Delta\chi^2 = 2403.43$, $\Delta df = 17$ i.e. $\Delta\chi^2/\Delta df = 141.37$, $p < .05$) indicated that common method bias may not be a serious problem in this study (Podsakoff et al., 2003).

Confirmatory factor analysis (CFA) using maximum likelihood estimation was conducted for the 10-factor measurement model. Results showed that the model fit the data reasonably well ($\chi^2_{417} = 640.67$, $p < .05$, RMSEA = 0.055, CFI = 0.94, NNFI = 0.95). Standardized loadings of the items ranged from 0.64 to 0.94; all were significant. Average variance extracted (AVE) by the underlying factors ranged from 0.63 (stability attribution) to 0.82 (private complaining). The squared correlations (Φ^2) among the underlying factors ranged from 0.00 (between locus attribution and desire for retaliation) to 0.17 (between public complaining and retaliation). These results collectively indicate that the measures exhibit convergent and discriminant validity (Fornell & Larcker, 1981). CFA results are shown in Table 1. Table 2 presents the descriptive statistics and internal consistency reliabilities (coefficient alphas) of composite scores of all constructs used.

Hypotheses testing

The model presented in Figure 1 was estimated via structural equation modelling using AMOS 20 to test the hypothesized relationships. Overall model-fit statistics have revealed that our proposed model fits the data well ($\chi^2_{430} = 675.39$, $p < .05$, RMSEA = 0.057, CFI =

Table 1. Confirmatory factor analysis of items and measurement properties of the scales.

Scale items	Standardized loadings	t-values
Perceived unfairness ($\Phi^2 = .00{-}.10$) (AVE = 0.79)		
In the hotel encounter you experienced, please indicate whether you were		
Treated unfairly	0.91	*
Treated wrongly	0.87	13.48
Unfairly dealt	0.90	15.16
Desire for retaliation ($\Phi^2 = .00{-}.17$) (AVE = 0.72)		
In the hotel encounter you experienced indicate whether you want to		
Do something bad to the hotel.	0.85	*
Take actions to get the hotel in trouble.	0.88	16.04
Cause inconvenience to the hotel.	0.90	16.52
Punish the hotel in some way.	0.88	15.85
Get even with the hotel.	0.73	11.58
Likelihood of firm response ($\Phi^2 = .00{-}.08$) (AVE = 0.75)		
In the hotel encounter you experienced, indicate to what extent you believe that		
The hotel will take appropriate action, if you complain about your negative experience.	0.81	*
The hotel will take appropriate action and give better service in the future, if you complain about your negative experience to the firm.	0.93	14.24
The hotel will give better service in the future and this will also benefit other consumers, if you complain about your negative experience to the hotel.	0.86	13.31
Stability attribution ($\Phi^2 = .00{-}.08$) (AVE = 0.63)		
In the hotel encounter you experienced, indicate to what extent you believe that		
This type of problem consistently happens in hotel industry.	0.89	*
This type of problem is likely to appear very frequently in the hotel industry.	0.83	11.10
This type of problem is likely to reoccur in the hotel industry.	0.64	8.87
Locus attribution ($\Phi^2 = .00{-}.09$) (AVE = 0.75)		
In the hotel encounter you experienced, indicate to what extent you believe that		
You are the one to blame (not the hotel) for the cause of the problem	0.86	6.68
You are the one (not the hotel) who is responsible for what has happened.	0.88	*
Personal Identity (PI) ($\Phi^2 = .02{-}.09$) (AVE = 0.66)		
In the hotel encounter you experience, you would post your experience in the Facebook because		
It is important to you personally.	0.80	*
It is a personal choice you feel responsible for.	0.94	12.11
You are interested in helping yourself.	0.68	9.71
Public complaining ($\Phi^2 = .00{-}.17$) (AVE = 0.73)		
Following the hotel encounter, how likely you are to post your experience in the hotel's Facebook page)		
Very likely to post	0.89	*
Inclined to post	0.80	13.48
Definitely will	0.88	15.16
Private complaining ($\Phi^2 = .00{-}.08$) (AVE = 0.82)		
Following the hotel encounter, how likely you are to post your experience in your Facebook page)		
Very likely to post	0.93	*
Inclined to post	0.88	18.32
Definitely will	0.92	15.16

(Continued)

Table 1. Continued.

Scale items	Standardized loadings	t-values
Post-complaining satisfaction ($\Phi^2 = .00-.10$) (AVE = 0.69)		
Given the hotel encounter, Please rate your feelings		
Overall, how satisfied or dissatisfied did this experience leave you feeling?	0.75	*
How well did this service experience meet your needs?	0.88	11.68
Overall, I am very satisfied with this experience?	0.87	11.55
Customer loyalty ($\Phi^2 = .00-.09$) (AVE = 0.71)		
Following the hotel encounter you experience, you are likely to		
Visit the hotel in the future.	0.85	*
Recommend this hotel to others.	0.89	14.12
Consider this hotel in the coming years.	0.79	12.33
Model-fit Statistics: $\chi^2_{417} = 640.67$, p = .00, RMSEA = .055, CFI = .94, NNFI = .95		

Note: All the loadings were significant at the 0.01 level.
*Loadings were initially fixed to 1.0.

0.94, NNFI = 0.94). The model explains 29% of the variance in public complaining using social media, 29% of the variance in private complaining using social media, 9% of the variance in PCS, and 32% of the variance in customer loyalty (See Table 3).

All proposed hypotheses predict the relative impacts of the measured constructs on public complaining and private complaining intentions using social media. First, the relative path coefficients of each construct were examined. They were then compared to its effect on public complaining and private complaining. Further, to provide a rigorous test for the stated hypotheses, equality restrictions were sequentially imposed on the effect of each construct on either public complaining or private complaining to allow a comparison of each χ^2 value with the baseline ($\chi^2_{430} = 675.39$) model free of such restrictions.

H1 states that perceived unfairness would have a more positive influence on public complaining than private complaining using social media. Results show that unfairness has a more positive effect on public complaining ($\beta = 0.21$, $p < .05$) than private complaining ($\beta = 0.19$, $p < .05$). To test the significance of the difference between two coefficients, equality restrictions were sequentially imposed for the effect of perceived unfairness on public complaining ($\chi^2_{431} = 737.20$) and compared with the baseline ($\chi^2_{430} = 675.39$) for a model without such restrictions. A significant χ^2 difference between the constrained and unconstrained models ($\Delta\chi^2 = 61.81$, $\Delta df = 1$, $p < .05$) was found, providing strong support for *H1*.

H2 holds that desire for retaliation has a more positive influence on private complaining than public complaining using social media. As shown in Table 3, the effect of retaliation on public complaining is not significant ($\beta = 0.11$, $p = .18$), but the effect of retaliation on private complaining is significant ($\beta = 0.27$, $p < .05$), supporting *H2*. Moreover, a significant χ^2 difference between the constrained and unconstrained models ($\Delta\chi^2 = 41.81$, $\Delta df = 1$, $p < .05$) was found, providing strong support for *H2*.

H3, which posits that likelihood of firm response would have a more positive influence on public complaining than private e-complaining using social media, was supported; likelihood of firm response has a positive effect on public complaining ($\beta = 0.32$, $p < .05$) and a negative effect on private complaining ($\beta = -0.20$, $p < .05$). Additionally, a significant χ^2 difference was found between the constrained and unconstrained models ($\Delta\chi^2 = 26.11$, $\Delta df = 1$, $p < .05$), providing additional support for *H3*.

Table 2. Correlations, means, standard deviations, and reliabilities of the composite scales.

Variables	1	2	3	4	5	6	7	8	9	10
1. Perceived unfairness	**1.00**									
2. Desire for retaliation	0.22	**1.00**								
3. Likelihood of firm response	0.11	-0.27	**1.00**							
4. Stability attribution	0.12	0.15	0.13	**1.00**						
5. Locus attribution	-0.19	0.00	-0.05	0.13	**1.00**					
6. Personal identity	0.08	0.21	0.08	0.18	0.08	**1.00**				
7. Public complaining	0.23	0.41	-0.20	-0.20	0.15	0.29	**1.00**			
8. Private complaining	0.27	0.09	0.28	0.28	-0.18	0.26	0.05	**1.00**		
9. Post-complaining satisfaction	-0.31	-0.08	-0.03	-0.03	0.30	0.30	-0.19	-0.16	**1.00**	
10. Customer loyalty	-0.15	0.05	-0.04	-0.04	0.15	0.15	-0.18	-0.08	0.30	**1.00**
Composite Scale Scores										
Mean	4.64	3.31	4.99	4.05	2.33	4.10	4.17	4.01	3.22	2.75
Std dev.	1.57	1.65	1.53	1.20	1.47	1.31	1.73	1.62	1.25	1.39
Cronbach alpha (α)	0.92	0.94	0.90	0.79	0.86	0.88	0.93	0.89	0.87	0.88

Note: Correlations > [0.14] are significant at the .05 level.

Table 3. Tests of structural model and research hypotheses.

	Standardized estimate	t-value	R^2
Unfairness → Public complaining	.21	2.54	
Retaliation → Public complaining	.11	1.32	
Firm response → Public complaining	.32	3.80	
Stability attribution → Public complaining	−.07	−0.82	
Locus attribution → Public complaining	−.17	−2.35	
Personal identity → Public complaining	.28	3.62	.29
Unfairness → Private complaining	.19	2.41	
Retaliation → Private complaining	.27	3.34	
Firm response → Private complaining	−.20	−2.52	
Stability attribution → Private complaining	.00	0.50	
Locus attribution → Private complaining	.01	0.01	
Personal identity → Private complaining	.26	3.50	.29
Public complaining → Post-complaining satisfaction	−.20	−2.51	
Private complaining → Post-complaining satisfaction	−.21	−2.57	.09
Public complaining → Loyalty	.01	0.16	
Private complaining → Loyalty	−.07	−1.02	
Post-complaining satisfaction → Loyalty	.55	6.20	.32
Model-fit statistics ($\chi^2_{430} = 675.39$, $p = .00$, RMSEA = .057, CFI = .94, NNFI = .94)			

Notes: Results are presented based on composite scale scores. All linkages with t values > [2.0] are significant at the .05 level.

H4 suggests that stability attribution would have a greater positive influence on private complaining than public complaining. As indicated by the non-significant coefficients as presented in Table 3 (public complaining $\beta = -0.07$, $p = .40$; private complaining $\beta = 0.04$, $p = .62$), this hypothesis was not supported. *H5* stated that locus attribution will have a more negative influence on public complaining than private complaining using social media. The coefficients presented in Table 3 indicate that the effect of locus attribution is significantly and negatively related only to public complaining ($\beta = -0.17$, $p < .05$), providing support for *H5*. Further, a significant χ^2 difference between constrained and unconstrained models ($\Delta\chi^2 = 88.51$, $\Delta df = 1$, $p < .05$) was found, offering additional support for *H5*.

H6 predicts that personal identity would have a more positive influence on private complaining than public complaining using social media. This hypothesis was not supported because personal identity had a slightly greater impact on public complaining ($\beta = 0.28$, $p < .05$) than private complaining ($\beta = 0.26$, $p < .05$). To assess if these differences are significant, the equality restrictions for the effect of personal identity on private complaining ($\chi^2_{431} = 735.90$) were imposed and compared with the baseline ($\chi^2_{430} = 675.39$) for a model without such restrictions. There was a significant χ^2 difference between the constrained and unconstrained models ($\Delta\chi^2 = 60.51$, $\Delta df = 1$, $p < .05$), failing to support *H6*.

H7 stated that private complaining would have a stronger negative influence on PCS than public complaining using social media. The respective coefficients (public complaining $\beta = -0.20$, $p < .05$; private complaining $\beta = -0.21$, $p < .05$) were examined and the equality restrictions for the effect of public complaining on satisfaction ($\chi^2_{431} = 875.0$) were imposed and compared with the baseline ($\chi^2_{430} = 675.39$) unrestricted model. Results have revealed a significant χ^2 difference between the constrained and unconstrained models ($\Delta\chi^2 = 199.61$, $\Delta df = 1$, $p < .05$), providing strong support for *H7*.

H8 was not supported because neither public nor private complaining behaviours were directly related to loyalty (public complaining $\beta = 0.12$, $p = .87$; private complaining $\beta = -0.07$, $p = .31$). As expected, *H9* was supported because PCS was positively related to

Table 4. Results of hypotheses testing.

	Hypotheses	Results
H1	Perceived unfairness has a more positive effect on public complaining using social media than private complaining using social media.	Supported
H2	Desire for retaliation has a more positive effect on private complaining using social media than public complaining using social media.	Supported
H3	Likelihood of firm response has a more positive effect on public complaining using social media than private complaining using social media.	Supported
H4	Stability attribution has a more positive effect on private complaining using social media than public complaining using social media.	Not Supported
H5	Higher levels of locus attribution have a more negative effect on public complaining using social media than private complaining using social media.	Supported
H6	Higher levels of personal identity have a more positive effect on private complaining using social media than public complaining using social media.	Not Supported
H7	Higher levels of private complaining via social media have a more negative effect on post-complaining satisfaction than public complaining using social media.	Supported
H8	Higher levels of private complaining via social media have a more negative effect on loyalty than public complaining using social media.	Not Supported
H9	Post-complaining satisfaction is positively related to loyalty	Supported

loyalty ($\beta = 0.55$, $p < .05$). We have highlighted key findings by summarizing the hypotheses' results in Table 4.

Discussion and implications

General discussion

An effective complaint management process is important for restoring customer satisfaction and loyalty. It reduces negative word-of-mouth, increases profitability and competitiveness, and helps in stabilizing the relationship with the service provider (Kaltcheva et al., 2013). A central element of successful complaint management is that customers voice their negative experience to the service provider. However, customers often engage in private complaining with friends and acquaintances to vent their anger and feel better. The emergence of social media has resulted in customers complaining about their dissatisfaction to the service provider and others using this channel. However, few studies have addressed the choice of social media as a complaint channel for service failures' context.

The present study contributes to the literature on service failure and consumer behaviour by exploring the factors that motivate customers to engage in complaint behaviours using social media. Based on theories of justice, attribution, and self-categorization, this study contends that perceived unfairness, desire for revenge, perceived service image, locus attribution, stability attribution, and personal identity determine CCB using social media. Specially, we show that the different factors influence private and public complaining using social media. Further, it was observed that CCB occurring via social media channels have unfavourable post-complaining outcomes.

Support for *H1* reveals that the level of perceived unfairness in a service failure encounter has a more positive effect on public complaining than private complaining using social media. This finding contrasts with some earlier studies (Von der Heyde Fernandes & Pizzutti dos Santos, 2008) where dissatisfaction level had a greater impact on negative word-of-mouth than voice complaints. As noted, this may be because customers feel that private complaining using social media does not result in problem resolution. Further, customers may engage in private complaining when they experience a series of failures or severe failure followed by unsuccessful recovery (Grégoire & Fisher, 2006).

The findings supporting *H2* suggest that the retaliation motive influences the choice of CCB using social media. This finding corroborates previous research, which shows that a customer's belief that a service provider violating the fairness norm results in a perception of betrayal (Ward & Ostrom, 2006; Gelbrich, 2010). In such cases, retaliation motives may direct the customer to engage in private complaining rather than public complaining using social media to denigrate the service provider. By sharing their dissatisfaction with others, the customers may harm the firm's reputation and motivate others to avoid patronizing the service provider.

The support for *H3* indicates that when customers perceive a higher likelihood of firm response to their complaints, they are more likely to choose public complaining than private complaining. A firm's policies and procedures in providing compensation or redress for service failures create the expectations that complaining leads to higher likelihood of success. In such cases, dissatisfied customers are more likely to voice their complaints to the service provider than complain to others. This finding is consistent with those of Huppertz (2007), who found that a firm's complaint policy and subsequent actions effectively lead to expectations of service recovery.

Because stability of the service failure was not related to CCB using social media, *H4* was not supported. One possible explanation is that attribution of stability affects the behavioural outcomes through the expectations of relationship continuity (Hess et al., 2003). Hence, future research should examine how stability attributions may influence CCB. The support of *H5* suggests that when customers attribute the reasons for failure to self, they are less likely to complain to the service provider using social media. This is consistent with Machleit and Mantel's (2001) findings that customers complain to the service provider when they attribute a failure directly to the provider. Interestingly, we did not find support for *H6*, the relationship between personal identity and CCB (public complaining: $\beta = 0.28$; private complaining $\beta = 0.26$, $\Delta\chi^2 = 27.9$, $\Delta df = 1$, $p < .05$). While personal identity was found to be positively related to both public and private complaining behaviours using social media, findings indicate that personal identity is significantly related to public complaining rather than private complaining. Public complaining using social media may better display the customer's self than private complaining, in which friends and acquaintances are already very well aware of one's personal identity.

Findings related to *H7* and *H9* indicate that CCB using social media is detrimental to the service provider. Specifically, complaining behaviours were negatively related to PCS with private complaining resulting in lower satisfaction than public complaining using social media, a finding consistent with prior research findings (e.g. Singh & Pandya, 1991). Similarly, as shown by Swanson and Hsu (2011), PCS is positively related to loyalty. *H8* was not supported as CCB were not directly related to loyalty. Taken together, the findings reveal that the impact of CCB on loyalty is through the mediating role of satisfaction.

Theoretical implications

The study findings add to the existing literature in several ways. First, this research integrates theories of justice, attribution, and self-categorization and bridges the service failure literature to the self-concept literature by examining CCB in the social media realm. More importantly, this study provides empirical evidence for the claim that CCB on social media depend not only on justice and attribution perceptions, but also from the extent to which customers perceive themselves and others on social media platforms or groups. Thus, this study enriches the current understanding of CCB and provides insights into the determinants of such behaviours.

Secondly, there has been limited research on channel-specific CCB. Mattila and Wirtz (2004) showed that customer rely on remote channels such as email to vent their frustration, but use real-time interaction channels (e.g. face-to-face or the telephone) for seeking tangible compensation. However, as new channels such as social media have emerged, there has formed a need for further research on complaining behaviours via these new channels. Further, this study has extended the research of Mattila and Wirtz (2004) and addresses the recent calls for research on social media as complaint channels (Clark, 2013) by developing and empirically examining an integrated model of CCB using social media. The results improve our current understanding of CCB using social media.

Finally, this study builds on previous work on CCB by examining the relative role of situational and personal factors in determining public and private complaining behaviours using social media. Based on Day and Landon's (1977) taxonomy, this research proposes that when service failures occur, customers can choose to engage in public and/or private complaining using a social media platform. While prior studies have largely explored the various forms of complaining behaviours and different complaint channels, the present study contributes further to the literature by examining the role of perceived unfairness, firm responsiveness, retaliation, locus attribution, stability attribution, and personal identity in determining public complaining and private complaining via social media. This advances the work on factors that motivate CCB following a service failure.

Managerial implications

As social media platforms are increasingly used as complaint channels, customers can more easily interact directly with the service provider. Moreover, the public nature of social media permits other customers to read the complaints and this may possibly influence their opinion towards the service provider. With the service provider's reputation and future business at risk as a consequence, acknowledging and addressing customer complaints in an appropriate manner is imperative for customer satisfaction.

Many academicians and practitioners have suggested that service providers require a new skill set to effectively manage customer complaints on social media. For example, it is important to monitor social media and respond to customer complaints with appropriate web-based interventions. As a first step, the service provider should devote appropriate human and financial resources to carefully monitor and track the social media platforms for customer complaints. They can invest in various monitoring tools such as social mention, tweet reach, to track customer complaints on social media. Moreover, dedicated social media teams can be established to respond to customer complaints and provide customer support via social media.

When publically complaining through social media, customers are concerned primarily with seeking an effective redress for the service failure. As this form of complaining is determined by perceived unfairness, responsiveness, locus attribution, and personal identity, the service provider should quickly acknowledge via social media the negative experience encountered by the customer and focus on recovering the service failure. This is particularly important because other customers on social media are watching. Hence, a quick and effective response is critical (Davidow, 2003). In short, the service provider should focus on fixing the service problem quickly and effectively because lack or delay of action can create a major public crisis. For instance, the average response time of Jet Blue in responding to its customer queries and complaints is 15 minutes. Einwiller and Steilen (2014) suggested that 50% of complaints on social networking sites are usually addressed within an hour. Hence, we propose that one hour is a reasonable time for service firms to acknowledge customer complaints on social networking sites (Grégoire et al., 2014). Moreover, blaming the customer for the service failure may lead to further disaster.

The service provider should also appreciate the customer and recognize the relevance of bringing the complaint to their notice. This enhances a customer's self-concept among other customers in social media. After acknowledging the complaint, the service provider can address the customer problem either privately or publicly depending on the severity of the service failure. After a successful recovery, the provider can easily respond to the customer via social media and let community members know that the problem has been resolved.

In cases of private complaining, customers have little interest in redress and are motivated mainly by their desire for revenge. Because these customers do not direct their complaints to the service provider, it is up to the service provider to use monitoring and tracking tools to make contact with the dissatisfied customers. By identifying customer complaints quickly, the service provider can communicate its commitment and competency, thereby avoiding potential public embarrassment and other consequences on social media. In addressing the private complaints, the service provider should acknowledge the complaint and communicate politely that it was not aware of the problem. This is important to counteract the customer's (and others') perception of no-action by the service provider in resolving the problem. Further, a service provider should communicate the recovery strategy on social media and invite the customer to engage with the service provider directly. In summary, a service provider must have a strong system in place for effective tracking of and response to customer complaints on social media.

Limitations, future research, and conclusions

Although this study expands our knowledge on an important topic, limitations must be noted, and viable prospects for further research should be identified. First, because the data were collected at a single point in time, the directionality of relationships must be addressed with caution. A true test of the causal structure of our model would require longitudinal data. Second, the study utilized Facebook as the context for examining e-complaining behaviours using social media. Smith, Fischer, and Yongjian (2012) argued that because social media platforms differ in their architecture, cultures, and norms, user-generated content posted on these platforms varies. Consequently, future research should examine the nature of complaining behaviours on various social media platforms. Third, this study has employed a convenience sample of undergraduate and graduate students. Because students are heavy users of Facebook, they are representative of the population for social media research. Future research, however, should test our hypotheses with a more broadly representative non-student sample. Fourth, this study examined the role of antecedent factors on CCB using social media. Future research should build upon this work by examining the role of relationship quality (Hess et al., 2003) and recovery options (Weun et al., 2004) in complaining behaviours.

Acknowledgment

Research conducted at Taylor's University institution.

Disclosure statement

No potential conflict of interest was reported by the authors.

Funding

The research was funded by TRGS/ERFS/2/2013/TBS/002 grant for the first two authors.

References

Alexandrov, A., Lilly, B., & Babakus, E. (2013). The effects of social and self-motives on the intentions to share positive and negative word of mouth. *Journal of the Academy of Marketing Science*, *41*(5), 531–546.

Balaji, M. S., & Sarkar, A. (2013). Does successful recovery mitigate failure severity?: A study of the behavioral outcomes in Indian context. *International Journal of Emerging Markets*, *8*(1), 65–81.

Bearden, W. O., & Oliver, R. L. (1985). The role of public and private complaining in satisfaction with problem resolution. *Journal of Consumer Affairs*, *19*(2), 222–240.

Blodgett, J. G., Hill, D. J., & Tax, S. S. (1997). The effects of distributive, procedural, and interactional justice on postcomplaint behavior. *Journal of Retailing*, *73*(2), 185–210.

Blodgett, J. G., Wakefield, K. L., & Barnes, J. H. (1995). The effects of customer service on consumer complaining behavior. *Journal of Services Marketing*, *9*(4), 31–42.

Bodey, K. & Grace, D. (2007). Contrasting "complainers" with "non-complainers" on attitude toward complaining, propensity to complain, and key personality characteristics: A nomological look. *Psychology & Marketing*, *24*(7), 579–594.

Casado-Díaz, A. B., & Nicolau-Gonzálbez, J. L. (2009). Explaining consumer complaining behaviour in double deviation scenarios: The banking services. *The Service Industries Journal*, *29*(12), 1659–1668.

Chiu, S. P., Chou, H. W., & Chiu, C. M. (2013). The Antecedents of buyers' perceived justice in online markets. *Cyberpsychology, Behavior, and Social Networking*, *16*(7), 536–542.

Clark, J. (2013). Conceptualising social media as complaint channel. *Promotional Communications*, *1*(1), 105–124.

Costarelli, S. (2009). Intergroup threat and experienced affect: The distinct roles of causal attributions and in-group identification. *The Journal of Social Psychology*, *149*(3), 393–401.

Davidow, M. (2003). Organizational responses to customer complaints: What works and what doesn't. *Journal of Service Research*, *5*(3), 225–250.

Day, R. L., & Landon, L. E. (1977). Toward a theory of consumer complaining behavior. In A. Woodside, J. Sheth & P. Bennett (Eds.), *Consumer and industrial buying behavior* (pp. 425–437). Amsterdam: North-Holland.

Eberly, M. B., Holley, E. C., Johnson, M. D., & Mitchell, T. R. (2011). Beyond internal and external: A dyadic theory of relational attributions. *Academy of Management Review*, *36*(4), 731–753.

Einwiller, S. A., & Steilen, S. (2014). Handling complaints on social network sites – An analysis of complaints and complaint responses on Facebook and Twitter pages of large US companies. *Public Relations Review*. doi:10.1016/j.pubrev.2014.11.012

Fornell, C., & Larcker, D. F. (1981). Structural equation models with unobservable variables and measurement error: Algebra and statistics. *Journal of Marketing Research*, *18*(3), 382–388.

Garnefski, N., & Kraaij, V. (2007). The cognitive emotion regulation questionnaire. *European Journal of Psychological Assessment*, *23*(3), 141–149.

Gelbrich, K. (2010). Anger, frustration, and helplessness after service failure: Coping strategies and effective informational support. *Journal of the Academy of Marketing Science*, *38*(5), 567–585.

Glynn, M. A. (2008). Beyond constraint: How institutions enable identities. In R. Greenwood, C. Oliver, R. Suddaby & K. Sahlin-Andersson (Eds.), *The Sage handbook of organizational institutionalism* (pp. 413–430). London: Sage.

Greenberg, J. (1987). A taxonomy of organizational justice theories. *Academy of Management Review*, *12*(1), 9–22.

Grégoire, Y., & Fisher, R. J. (2006). The effects of relationship quality on customer retaliation. *Marketing Letters*, *17*(1), 31–46.

Grégoire, Y., Salle, A., & Tripp, T. M. (2014). Managing social media crises with your customers: The good, the bad, and the ugly. *Business Horizons*, *58*(2), 173–182.

Harris, K. E., Mohr, L. A., & Bernhardt, K. L. (2006). Online service failure, consumer attributions and expectations. *Journal of Services Marketing*, *20*(7), 453–458.

Harrison-Walker, J. L. (2001). E-complaining: A content analysis of an Internet complaint forum. *Journal of Services Marketing*, *15*(5), 397–412.

Haslam, S. A., Branscombe, N. R., & Bachmann, S. (2003). Why consumers rebel: Social identity and the etiology of adverse reactions to service failure. In S. A. Haslam, D. V. Knippenberg, M. J. Platow & N. Ellemers (Eds.), *Social identity at work: Developing theory for organizational practice* (pp. 293–309). New York, NY: Psychology Press.

Heider, F. (1958). *The psychology of interpersonal relations*. New York, NY: Wiley.

Hess, R. L., Ganesan, S., & Klein, N. M. (2003). Service failure and recovery: The impact of relationship factors on customer satisfaction. *Journal of the Academy of Marketing Science*, *31*(2), 127–145.

Hogg, M. A., & Terry, D. I. (2000). Social identity and self-categorization processes in organizational contexts. *Academy of Management Review*, *25*(1), 121–140.

Holloway, B. B., & Beatty, S. E. (2003). Service failure in online retailing: A recovery opportunity. *Journal of Service Research*, *6*(1), 92–105.

Hong, J. Y., & Lee, W. N. (2005). Consumer complaint behavior in the online environment. Web systems design and online consumer behavior. In Y. Gao (Ed.), *Web systems design and online consumer behavior* (pp. 90–105). Hershey, PA: Idea Group publishing.

Hornsey, M. J. (2008). Social identity theory and self-categorization theory: A historical review. *Social and Personality Psychology Compass*, *2*(1), 204–222.

Huppertz, J. W. (2007). Firms' complaint handling policies and consumer complaint voicing. *Journal of Consumer Marketing*, *24*(7), 428–437.

Kaltcheva, V. D., Winsor, R. D., & Parasuraman, A. (2013). Do customer relationships mitigate or amplify failure responses? *Journal of Business Research*, *66*(4), 525–532.

Kaplan, A. M., & Haenlein, M. (2010). Users of the world, unite! The challenges and opportunities of Social Media. *Business Horizons*, *53*(1), 59–68.

Kim, C., Kim, S., Im, S., & Shin, C. (2003). The effect of attitude and perception on consumer complaint intentions. *Journal of Consumer Marketing*, *20*(4), 352–371.

Lim, S., & Palacios-Marques, D. (2011). Culture and purpose of web 2.0 service adoption: A study in the USA, Korea and Spain. *The Service Industries Journal*, *31*(1), 123–131.

Logan, K. (2014). Why isn't everyone doing it? A comparison of Antecedents to following brands on Twitter and Facebook. *Journal of Interactive Advertising*, *14*(2), 60–72.

Machleit, K. A., & Mantel, S. P. (2001). Emotional response and shopping satisfaction: Moderating effects of shopper attributions. *Journal of Business Research*, *54*(2), 97–106.

Marquis, M., & Filiatrault, P. (2002). Understanding complaining responses through consumers' self-consciousness disposition. *Psychology & Marketing*, *19*(3), 267–292.

Mattila, A. S., & Ro, H. (2008). Discrete negative emotions and customer dissatisfaction responses in a casual restaurant setting. *Journal of Hospitality & Tourism Research*, *32*(1), 89–107.

Mattila, A. S., & Wirtz, J. (2004). Consumer complaining to firms: The determinants of channel choice. *Journal of Services Marketing*, *18*(2), 147–155.

McColl-Kennedy, J. R., & Sparks, B. A. (2003). Application of fairness theory to service failures and service recovery. *Journal of Service Research*, *5*(3), 251–266.

McCollough, M. A., Berry, L. L., & Yadav, M. S. (2000). An empirical investigation of customer satisfaction after service failure and recovery. *Journal of Service Research*, *3*(2), 121–137.

Oliver, R. L., & Swan, J. E. (1989). Equity and disconfirmation perceptions as influences on merchant and product satisfaction. *Journal of Consumer Research*, *16*(3), 372–383.

Orsingher, C., Valentini, S., & de Angelis, M. (2010). A meta-analysis of satisfaction with complaint handling in services. *Journal of the Academy of Marketing Science*, *38*(2), 169–186.

Park, S. G., Kim, K., & O'Neill, M. (2014). Complaint behavior intentions and expectation of service recovery in individualistic and collectivistic cultures. *International Journal of Culture, Tourism and Hospitality Research*, *8*(3), 255–271.

Podsakoff, P. M., MacKenzie, S. B., Lee, J., & Podsakoff, N. P. (2003). Common method biases in behavioral research: A critical review of the literature and recommended remedies. *Journal of Applied Psychology, 88*(5), 879–903.

Rothenberger, S., Grewal, D., & Iyer, G. R. (2008). Understanding the role of complaint handling on consumer loyalty in service relationships. *Journal of Relationship Marketing, 7*(4), 359–376.

Settles, I. H., Jellison, W. A., & Pratt-Hyatt, J. S. (2009). Identification with multiple social groups: The moderating role of identity change over time among women-scientists. *Journal of Research in Personality, 43*(5), 856–867.

Sharma, P., Marshall, R., Alan Reday, P., & Na, W. (2010). Complainers versus non-complainers: A multi-national investigation of individual and situational influences on customer complaint behaviour. *Journal of Marketing Management, 26*(1–2), 163–180.

Singh, J., & Pandya, S. (1991). Exploring the effects of consumers' dissatisfaction level on complaint behaviours. *European Journal of Marketing, 25*(9), 7–21.

Smith, A. N., Fischer, E., & Yongjian, C. (2012). How does brand-related user-generated content differ across YouTube, Facebook, and Twitter? *Journal of Interactive Marketing, 26*(2), 102–113.

Sparks, B. A., & Browning, V. (2010). Complaining in cyberspace: The motives and forms of hotel guests' complaints online. *Journal of Hospitality Marketing & Management, 19*(7), 797–818.

Swanson, S. R., & Hsu, M. K. (2011). The effect of recovery locus attributions and service failure severity on word-of-mouth and repurchase behaviors in the hospitality industry. *Journal of Hospitality & Tourism Research, 35*(4), 511–529.

Swanson, S. R., Frankel, R., Sagan, M., & Johansen, D. L. (2011). Private and public voice: Exploring cultural influence. *Managing Service Quality, 21*(3), 216–239.

Tax, S. S., Brown, S. W., & Chandrashekaran, M. (1998). Customer evaluations of service complaint experiences: Implications for relationship marketing. *The Journal of Marketing, 62*(2), 60–76.

Tojib, D., & Khajehzadeh, S. (2014). The role of meta-perceptions in customer complaining behavior. *European Journal of Marketing, 48*(7/8), 1536–1556.

Tripp, T. M., & Grégoire, Y. (2011). When unhappy customers strike back on the Internet. *MIT Sloan Management Review, 52*(3), 37–44.

Turner, J. C. (1985). Social categorization and the self-concept: A social cognitive theory of group behavior. In E. J. Lawler (Ed.), *Advances in group processes: Theory and research* (pp. 77–122). Greenwich, CT: Jai Press.

Tyler, T. R., & Blader, S. L. (2003). The group engagement model: Procedural justice, social identity, and cooperative behavior. *Personality and Social Psychology Review, 7*(4), 349–361.

Van Stekelenburg, J., & Klandermans, B. (2013). The social psychology of protest. *Current Sociology, 61*(5–6), 886–905.

Von der Heyde Fernandes, D., & Pizzutti dos Santos, C. (2008). The antecedents of the consumer complaining behavior (CCB). *Advances in Consumer Research, 35*, 584–592.

Wan, L. C. (2013). Culture's impact on consumer complaining responses to embarrassing service failure. *Journal of Business Research, 66*(3), 298–305.

Wang, S., & Huff, L. C. (2007). Explaining buyers' responses to sellers' violation of trust. *European Journal of Marketing, 41*(9/10), 1033–1052.

Ward, J. C., & Ostrom, A. L. (2006). Complaining to the masses: The role of protest framing in customer-created complaint web sites. *Journal of Consumer Research, 33*(2), 220–230.

Weary, G., & Arkin, R. M. (1981). Attributional self-presentation. In J. H. Harvey, W. J. Ickes, & R. I. Kidd (Eds.), *New directions in attribution research* (Vol. 3, pp. 223–246). Hillsdale, NJ: Erlbaum.

Weun, S., Beatty, S. E., & Jones, M. A. (2004). The impact of service failure severity on service recovery evaluations and post-recovery relationships. *Journal of Services Marketing, 18*(2), 133–146.

Wu, I. L., & Huang, C. Y. (2015). Analysing complaint intentions in online shopping: The antecedents of justice and technology use and the mediator of customer satisfaction. *Behaviour & Information Technology, 34*(1), 69–80.

Zboja, J. J., & Voorhees, C. M. (2006). The impact of brand trust and satisfaction on retailer repurchase intentions. *Journal of Services Marketing, 20*(6), 381–390.

Appendix

At the busiest time of the summer season, you made a reservation for a hotel that is part of a domestic hotel chain at a popular tourist location. You have decided to spend a short vacation at this hotel with your family members because it offers many features including a health club, swimming pool, quality restaurants and a reputation for giving a special attention to its guests. This will be your first time there. When you arrive at the hotel with your family on the day of reservation, you found that the room would not be available due to a registration mistake. The receptionist apologies for the inconvenience but informs that the room would be available the next day and gives you a confirmation number for your reservation with a 3 p.m. check-in time the next day. Naturally, you feel very disappointed.

Marketing technology for adoption by small business

Philip Alford and Stephen John Page

Faculty of Management, Bournemouth University, Fern Barrow, Poole, UK

The adoption of technology for marketing is essential for the survival of small businesses and yet little is understood about owner-manager practice in this area. This paper aims to address that gap through a qualitative study of 24 owner-managed small businesses operating in the visitor economy. It found that there was a strong appetite for the adoption of technology for marketing and a clear recognition of its opportunities particularly related to how it could create a stronger market orientation and more agile marketing, adhering to the principles of effectual reasoning. However, the ability to take advantage of these opportunities was constrained by a lack of knowledge and in particular an inability to measure the return on investment. While the wider implications of the study are limited by the niche sample, a planning model for the adoption of technology for marketing is presented which can be tested through future research.

Introduction

The Internet has the potential to transform small and medium enterprise (SME) marketing in a number of areas including customisation, customer relationship marketing, access to new markets, business-to-business collaboration, co-creation of the product with customers and improving internal efficiency (Ansari & Mela, 2003; Barnes et al., 2012; Harrigan et al., 2010; Harris & Rae, 2009). Industry sources such as the consulting firm McKinsey noted in 2011 that SMEs who have a strong Web presence grow twice as quickly as those who have no or minimal presence.

According to Kim, Lee, and Lee (2011), 'all firms that utilise the unique value of Web 2.0 through superior management skills, innovation, and business process reengineering are likely to enjoy sustainable competitive advantage' (p. 158). However, despite the importance, the evidence suggests that SME adoption of the Internet is limited. Only one-third of SMEs in the UK have a digital presence according to trade organisations such as the Federation of Small Business, with established businesses finding the challenge of integrating fast moving technology into their business 'daunting', while other industry sources (i.e. Sage UK, the accountancy software firm), found that in 2014 three quarters of British SMEs do not use social media to engage with local consumers. It is clear that SMEs are facing barriers, which are preventing them from fully harnessing the potential of the Internet for marketing (Kim et al., 2011).

This study focuses on the owner-managers of small and micro enterprises (1–49 employees as defined by the European Commission), with evidence suggesting that this subgroup warrant unique investigation within the SME population. There were 5 million micro enterprises in the UK in 2014, accounting for 96% of all businesses and for 92.4% of all enterprises in the 28 EU member states in the non-financial business sector. The Federation of Small Business recommends that the Office of National Statistics extend its 'Information and Communications Technology (ICT) Activity of UK Businesses' study to include micro enterprises. The European Commission has set new financial thresholds in its definition of small businesses to begin to encourage Europe-wide measures to address the problems micro enterprises face. The paper's principal contribution to theory in the field of services research lies in the development of a planning model for the adoption of technology for marketing by small businesses. This model, which can be tested through future research, is derived through the research undertaken for this study.

Jones, Simmons, Packham, Beynon-Davies, and Pickernell (2014) argue that because of size and resource issues, micro enterprises cannot be studied in the same vein as larger SMEs and they 'would be expected to exhibit unique attitudes and strategic responses towards ICT adoption' (2014, p. 289). Barnes et al. (2012) remind us that small businesses are not little big businesses. Simmons, Armstrong, and Durkin (2008) single out the role of the owner as having more significance in a micro enterprise than in their larger counterparts. Citing Standing, Vasudavan, and Borberly (1998), Ritchie and Brindley (2005) point out that the rate of technology adoption in a micro enterprise is highly dependent on the passion and interest that the owner-manager has for technology. However, despite their economic significance and their unique characteristics, in particular the central role of the owner, there is a limited body of research into how owner-managers are adopting technology for marketing and relatively little empirical research into the attitudes of owners in this area.

In order to address these gaps in knowledge, the objectives of this exploratory study are as follows: firstly, to review the current literature on the adoption of technology for marketing by small businesses and owner-managed businesses in particular. Secondly, to empirically investigate marketing technology for adoption by small business. Thirdly, to understand the barriers which prevent adoption of technology for marketing. Fourthly, to identify ways in which small businesses can be more effectively supported in their use of technology for marketing and to identify future directions in this field so as to test the model presented in this study.

In order to address these objectives, this study explored the adoption of technology for marketing by 24 small owner-managed tourism-related businesses operating in the visitor economy. Tourism has been identified as a sector worthy of investigation owing to its reputation for being 'dynamic in website adoption' (Simmons, Armstrong, & Durkin, 2011, p. 552), although this study goes beyond a website focus. Tourism-related small businesses are heavily reliant on the Internet in a market where customer relationships are increasingly mediated online. These businesses are challenged with promoting an intangible service where the information they provide online is often the only tangible representation of the service that the customer will experience at the point of decision making and purchase. The seasonality of demand and the perishable nature of the product compound these challenges.

The paper is structured as follows. Firstly, the literature is reviewed to identify the issues, which small business owner-managers face, related to technology for marketing adoption. Secondly, the methodology is described. Thirdly, the findings are presented and then discussed within the context of the literature. Fourthly, an empirically informed model is presented, to guide the adoption of technology for marketing by small business owner-managers. Finally, conclusions are reached and areas for future research are suggested.

Small business adoption of technology for marketing

There are a growing number of studies in the literature related to the adoption of technology for marketing by small businesses (Bulearca & Bulearca, 2010; Durkin, McGowan, & McKeown, 2013; Harris & Rae, 2009; Kim et al., 2011; McGowan & Durkin, 2002; Simmons et al., 2008, 2011) and some focusing on micro enterprises in particular (Barnes et al., 2012; de Berranger, Tucker, & Jones, 2001; Bharati & Chaudhury, 2006; Dandridge & Levenburg, 2000; Jones et al., 2014; Rhodes, 2009; Wolcott, Kamal, & Qureshi, 2008). The following review of this literature is structured according to themes, which essentially represent the main issues impacting on the adoption of technology for marketing by small business owner-managers.

The first of these relates to the highly influential role played by the owner: 'in order to understand micro-enterprises' ICT adoption, researchers need to appreciate the attitudes of owner/managers which determine their strategic response' (Simmons et al., 2008, cited in Jones et al., 2014, p. 286). Fillis and Wagner (2005) identify the level of entrepreneurial orientation of the key decision maker as instrumental in determining the adoption of e-business. In addition to attitudes and orientation, McGowan and Durkin (2002), in their study of entrepreneurial small firms, found that the competency of the owner in the areas of vision, value, technical ability and control, coupled with the extent to which customers, collaborators, and competitors are using the Internet, influences adoption.

Secondly, and related to the first theme, there is more likely to be successful adoption of technology where the owner perceives it as an enabler of the business and can see its longer term benefits. Simmons et al. (2011) refer to this as having an eVision, which is directly dependent on the owner having a clear perception of the benefits. Martin and Matlay (2003) found that small businesses could see the benefits of websites, viewing them as 'essential marketing tools' that provide access to new customers, build stronger links with existing customers, access niche markets and 'change customers perceptions of the firm's products or services' (Martin & Matlay, 2003, p. 23). Jones et al. (2014) found that micro enterprises were more likely to adopt technology where they could see a close fit with their business model. The term Web 2.0, originally coined by O'Reilly (2007), is often used (Barnes et al., 2012; Hinchcliffe, 2010; Kim et al., 2011) to denote a shift from 'a business-centred to a user-centred model' (Kim et al., 2011, p. 157).

The third theme concerns the link between market orientation and technology adoption. While there is a link between a strong marketing orientation and business performance (Kara, Spillan, & DeShields, 2005), there is less clarity about the role played by technology in enabling that orientation in small businesses. Peltier, Zhao, and Schibrowsky (2012) suggest that the link between marketing orientation and the successful adoption of new technology is unclear. However, a number of studies have concluded there is a positive link between market orientation and technology adoption among small businesses (Coviello, Milley, & Marcolin, 2001; Polo Peña, Frías Jamilena, & Rodriquez Molina, 2011). Simmons et al. (2011) draw on Pelham and Wilson's (1996) small business marketing orientation scale to help to identify the qualities of a market-oriented business (i.e. targeted marketing; understanding of customer needs; creating customer value; response time to negative customer feedback; sharing customer information across the organisation; and internal strengths and weaknesses and competitor analysis) and market orientation is central to their website optimisation model. However, they acknowledge that the link between market orientation and developing an online value proposition 'is an area that requires a more developed understanding from the small business perspective' (Simmons et al., 2011, p. 537).

The fourth theme pertains to the barriers to adoption of technology for marketing, including resource factors (e.g. time and money) and these limitations are well documented in the small business literature (Barnes et al., 2012; Street & Cameron, 2007; Wymer & Regan, 2005). Perhaps more troublesome is the proposition, as Bharadwaj and Soni (2007) found, that small business disengagement with e-commerce was a perception of it lacking strategic importance. A further potential barrier is the challenge that small businesses face in integrating 'traditional marketing practices with new web-based opportunities' (Simmons et al., 2008, p. 364) and the opportunity cost whereby small businesses may view investment in online marketing at the expense of more traditional marketing.

A related barrier is the limited extent to which small businesses experiment with technology in view of the resource implications (Simmons et al., 2008, p. 358). This has a knock-on effect insomuch that if small businesses fail to test how the adoption of technology can add value to their marketing, they are less likely to identify the full costs and benefits and consequently be less inclined to incorporate it into their overall business strategy. This failure to test and experiment is compounded by the lack of measurement; for example, Sellitto, Andrew, and Burgess (2003) found that while small businesses had a general purpose for their online presence, they had no specific goals for their website. A lack of goal setting and measurement of the impact of the Internet on marketing may breed further 'scepticism' (Kim et al., 2011, p. 170).

Fifthly, small businesses tend to opt for a just doing approach rather than for formal planning (Chaston, Badger, Mangles, & Sadler-Smith, 2003; Jones, Hecker, & Holland, 2003). This has led recently to studies exploring the way in which effectuation might help to provide insights into technology adoption by small businesses (Fischer & Reuber, 2011). 'Effectual logic is the name given to heuristics used by expert entrepreneurs in new venture creation' (Read, Sarasvathy, Dew, Wiltbank, & Ohisson, 2011, p. 7). The relevance of effectuation, which 'evokes creative and transformative tactics' (Read et al., 2011, p. 7), as a lens through which to view technology for marketing adoption is of particular interest in the context of digital media where the opportunities for customer engagement are significantly greater than they were through more traditional media. Furthermore, this engagement is highly measurable, whereby almost all forms of online customer engagement can be tracked and analysed and the insights used to feed back into marketing activity. The contribution of effectual logic to the field of small business use of technology for marketing is discussed later in the paper after the findings are presented.

A sixth and final theme is the growing number of studies focusing on the adoption of the wider online marketing mix and Web 2.0 applications by SMEs (Barnes et al., 2012; Bulearca & Bulearca, 2010; Durkin et al., 2013; Harris & Rae, 2009; Kim et al., 2011) as opposed to only the website which traditionally has been the focus. This transition from a website-specific focus to an expanded range of technologies such as mobile apps, application programme interface, social networks and analytics has created a complex landscape, which is increasingly challenging for small business owner-managers to navigate.

Obtaining a clear view of the costs and benefits in this new era and setting measurable goals as part of a wider strategy becomes considerably more difficult. This is in part because in a multichannel environment, where the creation and distribution of content that influences the customer is increasingly created and shared by other customers, many elements of online marketing are beyond the control of the business. This 'earned media' (Chaffey &

Ellis-Chadwick, 2012), also referred to as 'user-generated content' (Kim et al., 2011), is growing in influence (Dhar & Chang, 2009). Qualman (2013) coined the term Socialnomics to capture the power and influence of this type of content. The magnitude of this change has led the Marketing Science Institute to argue that a major shift is needed to embrace these new themes. Despite increasing coverage in the literature relating to digital marketing by small businesses, there remains a lack of conceptual and empirical studies (Kim et al., 2011) to explain how they make sense of this turbulent technological environment (Peltier et al., 2012).

Methodology

In order to address the objectives of the study and to provide more insights into the themes discussed above, a qualitative methodology was deployed and data obtained from a convenience sample of 24 owner-managed tourism-related enterprises (Table 1 for a profile of the owner-managed businesses). A qualitative methodology is appropriate considering the exploratory nature of the study. The enterprises that comprised the sample were part of a wider 60-strong SME cohort, comprising both owner-managers and managers, that applied to be part of a university-led project investigating the adoption of technology for marketing; therefore, this was a group that was using technology to support their marketing but were interested in exploring opportunities in more depth, thereby providing a rich information source.

The research design for the entire project included three phases: the first was exploratory designed to investigate levels of adoption of technology for marketing; the second was

Table 1. Profile of owner-managed businesses.

Case	Business type	Employees FT and PT
A	Tourist attraction	25
B	Bed & breakfast	1
C	Tourist attraction	4
D	Holiday accommodation/property agency	20
E	Hotel	9
F	Bed & breakfast	1
G	Sightseeing tours	30
H	Self-catering accommodation	10
I	Wine education and training	3
J	Retail/restaurant/online sales/public garden	30
K	Hotel & restaurant	35
L	Professional conference organisers	2
M	Sightseeing tours	1
N	Adventure sports/outdoor education	40
O	Bed & breakfast	1
P	Outdoor activities	5
Q	Self-catering accommodation	1
R	Self-catering accommodation	1
S	Corporate yacht charters	17
T	Tourist attraction	45
U	Outdoor activities	6
V	Hotel	25
W	Self-catering accommodation	1
X	Visitor accommodation and activities	1

an action research phase whereby the participating businesses undertook online marketing campaigns; the final phase reflected on the outcomes and the learning that took place. This paper reports on the first phase of the project where the data were gathered through a number of methods: firstly, each business completed an application form, which asked for information relating to them as individuals (e.g. reasons for applying, previous experience, personal professional development, and attitudes to technology) and their business (size, current adoption of technology for marketing, digital marketing goals, key strengths and weaknesses).

Secondly, participants met twice in smaller clusters of 10 to discuss the challenges and opportunities surrounding technology for marketing and their attitudes and opinions. The first half-day cluster meeting was facilitated at the university and participants were asked to outline their priorities for digital marketing and the challenges they faced. This enabled a wider discussion and at times the facilitators would prompt and probe along the themes identified in the previous section of this paper. The second meeting was self-organised by each cluster, lasting on average 90 minutes, and built on the themes and ideas discussed at the first meeting. Participants were tasked with refining their digital marketing priorities in order to provide a focus for the meeting and to lead into the second action research phase.

A number of measures were built into the research design to enhance trustworthiness and authenticity of the study. Firstly, the first group discussion was facilitated by the project lead at the university and an external consultant who specialises in marketing for small businesses. This ensured a practitioner perspective that was important in communicating the purpose of the study in terms that owner-managers could relate to. It also provided a quality check on data collection and interpretation and the identification of key themes. We were aware of the importance of creating a space that would allow the participants to openly share their experiences, concerns and objectives relating to technology adoption. The facilitation by a consultant experienced in working with small business owners helped to ensure a level of empathy which proved effective in encouraging owners to express their viewpoints. The interaction that took place among the owners was also instrumental in developing a free flow of conversation and additional insights that may not have been possible in an individual interview. Secondly, the first group discussion was captured on video, which enabled the research team to capture important visual indicators. Thirdly, a member of the research team attended the cluster meetings and used participant observation as a quality check. Fourthly, additional rigour was added through the feedback which businesses provided via blog posts on the website supporting the study and through a private discussion forum on LinkedIn. Fifthly, the group discussion at the cluster meeting in particular featured a high level of peer-to-peer interaction, which facilitated clarification.

Data management and analysis

The application forms and transcripts of the group discussions were uploaded to NVivo, a qualitative data analysis package that enables coding, querying, visualising, and reporting on the data (Bazeley & Jackson, 2013). A subset of the data was established for the 24 owner-managers, which forms the basis for the analysis in this paper. The text was coded and analysed 'to form concepts (thematic analysis) and identify relationships (semantic analysis) between concepts' (Verreynne, Parker, & Wilson, 2013, p. 410).

The second step was the process of 'going on with coding' (Bazeley & Jackson, 2013, p. 95) whereby a structured coding system (thematic framework) was created where the parent nodes represented themes central to the study and the child nodes, sub-themes. This paring-down process provided 'conceptual clarity' (Bazeley & Jackson, 2013, p. 97). While the thematic framework was informed by the literature (deductive reasoning), the process also enabled the identification of concepts that emerged from the data (inductive reasoning) and these are discussed as follows.

Findings

A number of themes and sub-themes emerge from the research with the businesses, which provide valuable insights into owner-manager adoption of technology for marketing. The dominant theme in terms of the amount of text references (denoted in parentheses) coded to it, is market orientation (83), followed by strategy and vision (56), barriers to technology for marketing adoption (46), measurement (39), owner profile (20), collaboration (17), and benefits (12).

The market orientation sub-themes are as follows: Unique Selling Proposition (USP) (28), targeting (19), customer insight (12), customer conversation (10), linking channel to customer (9), customer retention (6), product–market match (6), customer database (5), and customer journey (3).

Three of these sub-themes are listed by Pelham and Wilson (1996) in their market orientation framework, namely target marketing, creating customer value (coded to USP), and understanding customer needs (coded to customer insight). However, three further themes which Pelham and Wilson identified were not referred to by owners (i.e. response time to negative customer feedback, sharing customer information across the organisation and competitor analysis). This suggests a number of missed opportunities for owner-managers as social media sites such as Twitter and Facebook can provide highly responsive customer service channels. The transparency of the Internet enables competitor analysis, for example, Facebook Insights allows Facebook Page owners to view analytics of five other pages. In addition, the increasing range of free or low-cost applications in the cloud facilitates the sharing of information between staff, which can enable the more flexible working practices characteristic of small businesses (e.g. homeworking, seasonal staff).

The research with the business owners identified a number of additional sub-themes, which indicates that they have a level of market orientation and, importantly, demonstrates an awareness of how technology can facilitate that. One example is customer conversation:

> I think the idea of a conversation with people through digital … I'm finding more and more now Facebook is a conversation. (Jackie, owner-manager, Tourist attraction, 45 employees)

Having a conversation with customers did not exist as a marketing concept five years ago and is a product of the Web 2.0 era, particularly the interactivity afforded by social media, which in addition to communication also provides the opportunity to gain customer insight:

> So for me, Twitter's not just a one-way thing, I'm learning a lot about our market through it as well. (Karen, owner-manager, Self-catering accommodation, 1 employee)

Simultaneous reference to more than one market orientation theme is a pattern that emerged from the data and demonstrates a sophisticated level of owner-manager knowledge and an understanding of how the adoption of technology can facilitate that orientation:

> Tim (owner/manager, Sightseeing tours, 30 employees) and I are obviously in a similar business, but very different, but I think the word experience, which is a word I've tried to use, we've got the semi retired market and we've also got the dynamic hen and stag market as well, and I think trying to get through to both those markets which are both very fertile I

think is an interesting ... and that's where I think Facebook, and that's where I'll be focusing on trying to get those two markets addressed. (Andy, owner-manager, Sightseeing tours, 1 employee)

The owners demonstrated an awareness of the importance of communicating their USP online but were challenged as to how to achieve this:

We are a very visual business – plants, flowers, food, growing vegetables so there is always a story to tell – how can we make the most of this on line and increase visitors and sales? (Patrick, owner-manager, Public garden, 30 employees)

This relates to one of the barriers to adoption of technology for marketing that emerged during this study, namely the difficulty owners perceive in creating online content, particularly in social media channels. The reference to storytelling in the quote above is an emerging trend (Black & Kelley, 2009; Tussyadiah & Fesenmaier, 2008) and the opportunity to utilise customers' stories which are readily available online may help to provide small business owners with ideas for communicating their USP online.

The concept of the online customer journey emerged from the data and while it demonstrates an awareness of customer behaviour on the one hand, the growing complexity of that journey represents a significant issue for owner-managers:

We had this conversation as well, we started to talk about measuring, and the challenges in the good old days, someone could ring up your staff and do a holiday booking and say where did you find us, and they'd say oh we saw your advert in whatever newspaper, oh great, there you go. Since Google and websites and things like that it's the routes they take to actually end up on your site. I completely agree with what you guys were saying about trying to find that route of how they made their journey actually into your company is almost impossible nowadays. (Des, owner-manager, Holiday property agency, 20 employees)

The nostalgic reference to 'the good old days' when traditional advertising was relatively easy to measure coupled with the reference to the 'impossibility' of finding the route they took to the company underlines the challenges which owner-managers face in attributing the business they gain from digital marketing and accurately measuring return on investment (ROI).

Measurement emerged as a strong theme and represents both an aspiration and frustration for owner-managers:

In our case most of our enquiries still come over the phone, even though people go to the website first. This makes tracking conversions rates tricky and would like to find ways of better monitoring our online impact and some tighter goals to aim for. (Sam, owner-manager, Outdoor activities, 5 employees)

But I think it's really understanding the measurement of that and what our engagement is, and ultimately what that puts on the bottom line. I think that's a key message for us. We do lots of things and I haven't got the slightest idea what works. (Jackie, owner-manager, Tourist attraction, 45 employees)

One of the advantages of online marketing is the ability to measure customer interactions more accurately than is possible with traditional marketing practices (e.g. print, television, and radio). Most online marketing channels come with their own analytical tools that allow a range of important marketing metrics to be captured. However, a combination of lack of time and knowledge of how to use these tools and integrate the information they provide into their marketing represents a significant barrier to further adoption of technology for marketing by owner-managers.

However, there was an appetite among owner-managers to test and learn and to experiment with online marketing channels and, equipped with the knowledge, to realise tangible benefits:

When it started, I think we spent about £70 a month on ad words and we were doing things like things to do in Bournemouth. But what was happening is, we'd watch them land on the page and with analytics you can see how many land on the page and then how many drop off. And then we were looking back at the data and analytics, so we were getting rid of key words, so we only end up with about five or six for each ad campaign. And now it's down to about £9. But we were picking up the right contacts. (Andy, owner-manager, Outdoor activities, 6 employees)

Owner-managers expressed more interest in measuring the impact of their current online marketing and obtaining a better focus rather than developing long-term strategy:

We were talking about putting a strategy plan together, but actually it's got to be more focused now, actually saying right, this month what we're trying to get out of it. Is it number of likes we're trying to get, is it number of bookings, and actually number of reviews at the end of it, and actually start to say right, this month we are focusing on this and that's got to be our goal, because otherwise you can get lost amongst it all. (Olivia, owner-manager, Hotel, 25 employees)

The goal of setting a longer term strategy or plan was referred to frequently in the application forms, however, subsequent discussions revealed that shorter term tactical issues were of more importance to owner-managers.

Discussion of findings

The findings indicate that small business owner-managers have a positive attitude and a real appetite for adopting technology for marketing, associating it with an opportunity to engage with customers at a deeper and more interactive level. There is clear recognition among owner-managers that the adoption of technology for marketing is beneficial although this did not extend to the level of recognising fit with their business model (Jones et al., 2014). In addition, there is an awareness of how the adoption of technology for marketing can facilitate a clear market orientation (Coviello et al., 2001).

The findings also reveal a number of Web 2.0 market orientation concepts (i.e. online customer journey and online conversations) that go beyond current frameworks (Pelham & Wilson, 1996), and offer potential for owner-managers to shift from 'a business-centred to a user-centred model' (Kim et al., 2011, p. 157). This is in line with the open and collaborative approach advocated by Harris and Rae (2009) using Web 2.0 tools, for example, Twitter, as a two-way channel for communicating and listening to the customer (Bulearca & Bulearca, 2010). However, there is also an opportunity for owner-managers to adopt technology to gain competitor insight and to share information across the organisation (Pelham & Wilson, 1996).

However, the ability for owner-managers to turn this positive view of adopting technology for marketing into a clear plan is hindered by a lack of technical competency (McGowan & Durkin, 2002), particularly the inability to set and measure goals (Sellitto et al., 2003). This acts as a barrier to realising the full benefits of the adoption of technology for marketing and consequently developing an eVision (Simmons et al., 2008). The data do not support the findings of Bharadwaj and Soni (2007) that small business disengagement with e-commerce was a perception of it lacking strategic importance. On the contrary, these owner-managers were very aware of the importance of technology-enabled marketing, explaining in large part their frustration at not being able to implement and measure it more fully.

It has been documented for some time that owner-managers tend to do rather than plan (Chaston et al., 2003). They adhere more to the principles of effectuation (Read et al., 2011), namely seeking to control and shape an unknown and unknowable future rather than to predict it. This is particularly relevant in the fast moving Web 2.0 landscape in which owner-managers must compete and where prediction is increasingly difficult. The owner-managers in this study are challenged by customer engagement online and find their inability to understand and control it stressful. However, with the requisite technical skillset, there are opportunities in this new landscape. The means (who I am, what I know, and who I know), which an entrepreneur has at his or her disposal, is the starting point in the effectual cycle (Sarasvathy, 2001). Online customer engagement generates a large amount of data that can provide the owner-manager with additional means (what they know, who they know), which enable them to set new goals (Fischer & Reuber, 2011). For example, at an operational level, customer feedback via social media can alert an owner-manager to opportunities for improving their service offer. At a more strategic level, digital marketing analytics can help an owner-manager to identify new customer segments to target, including goals to reach that segment and indicators to measure those goals.

On a business-to-business level, owner-managers in this study demonstrated a keen interest in forming partnerships and recognised that collaboration could provide an advantage for their business. While the relationships were initially formed face to face, a number of owner-managers in this study continued to interact with each other via Twitter after the project had finished. This highlights the opportunity for small business owner-managers to create clusters and to explore the co-creation of products and experiences with partners in that cluster. It is instructive that the strongest networks within the study were those comprising owner-managed micro firms, which, as indicated above, extended networking beyond the confines of the study through exchanges on Twitter. In one case two micro firms, a small bed & breakfast and a cycle hire company, discussed business collaboration in a face-to-face meeting outside the study. Micro firm owner-managers were particularly keen in learning from other participants' experiences in implementing technology, lending credence to the importance of peer-to-peer exchange. The role of the owner-manager is often an isolated one, particularly in a rural setting, and the opportunities to discuss technology adoption with their peer group are very limited.

However, while effectuation provides an interesting lens through which to view owner-manager adoption of technology for marketing, it does not equate to 'not planning'. Not having preset goals does not mean 'not having goals' and this study has demonstrated that goals are vital for the successful adoption and measurement of technology for marketing. However, in keeping with the principles of effectual reasoning, it is more appropriate to think in terms of shorter, more adaptive, planning cycles, informed by insight through measurement. A key implication of this research therefore is the requirement for small business owner-managers to improve their technical competency in order to realise the full benefits, particularly in terms of more agile and user-centred, data-driven marketing.

This study demonstrates that owner-managers require new skillsets in order to overcome the barriers to adopting technology for marketing. They mirror those in social media customer relationship marketing, including an ability to analyse data, measure ROI, integrate customer touch points and create engaging content (Malthouse, Haenlein, Skiera, Wege, & Zhang, 2013). Malthouse et al. (2013) identified the need to change the culture of the organisation and provide the workforce with the required skillset including data analysis (data scientists), interpretation, and business intelligence. This presents a daunting challenge but is imperative to address if micro firms are going to realise the benefits of more agile marketing. Owner-managers of micro firms tend to learn through doing, adhering to the principles of effectual entrepreneurship as opposed to formal planning. However, in today's complex digital environment that learning process is more difficult. The following section presents a model to facilitate a test-and-learn approach for micro firm marketing. The willingness of owner-managers to collaborate in and beyond this study presents an opportunity for them, collectively, to share the knowledge and insights they gain from testing this model within business clusters. Equally, collaboration presents an opportunity to enhance the service offer and develop a more unique experience for the customer.

Practical implications of the study

This section prescribes an empirically informed model (Figure 1) to guide owner-manager adoption of technology for marketing; it is divided into three interconnected phases, with each phase tied to the findings of the study.

In phase 1, owner-managers develop a clear market orientation through customer targeting and identifying their USP(s). This includes reading content that customers create about the business in order to obtain the customer's perspective on the value provided by the experience, accessing digital marketing channels analytics to understand consumer online behaviour and data mining of customer databases. This will help businesses to create customer personas which are used in human-centred design and marketing (Chaffey, 2010; Hisanabe, 2009) and go beyond traditional socio-demographic market segmentation methods that are being challenged in marketing (Valos, Ewing, & Powell, 2010). The findings of this study demonstrate that owner-managers are aware of the broad segments that they are targeting but lack deeper insights into online customer behaviour which is inhibiting their ability to communicate their unique propositions. The realisation among participants in this study of the need to embed analytics in their strategy can be harnessed to enable this phase of the model.

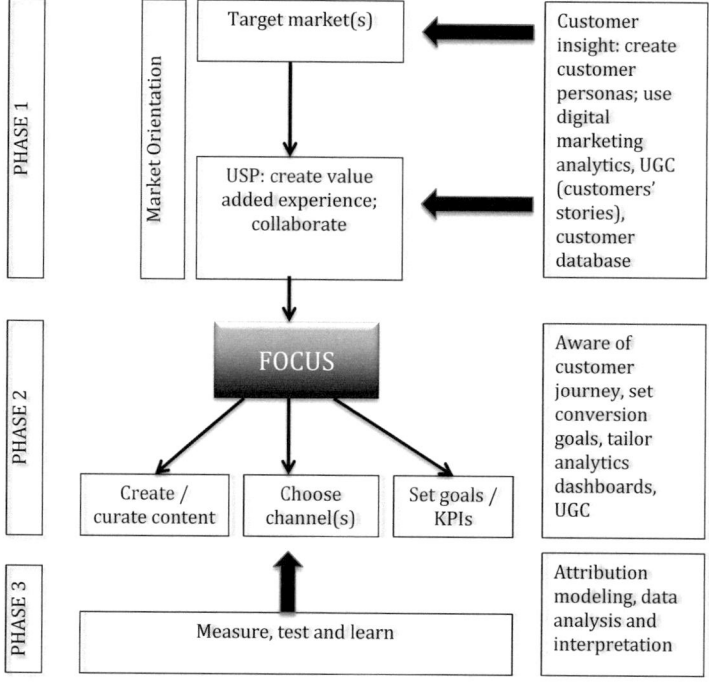

Figure 1. Technology for marketing adoption model for small businesses.

Phase 2 of the model provides owner-managers with a focus for customer acquisition and retention, building on the customer insight gained in Phase 1. However, this study suggests that owner-managers do not pay sufficient attention to engaging with repeat customers. This is a missed opportunity as it is less costly to retain existing customers and there are opportunities to engage more effectively with them based on customer knowledge (Harrigan et al., 2010; Payne & Frow, 2005; Sigala, 2006). While owner-managers in the study were broadly aware of the opportunities that social media offered for customer engagement, for example, through the concept of conversation, they were not clear on how to utilise and measure it. One of the participants quoted earlier (Patrick, owner-manager, public garden, 30 employees) referred to the difficulty he had in telling his story. However, there is an opportunity for small businesses to build a narrative from the perspective of their customers by curating the content that they create. For example, Storify (www.storify.com) provides a platform for small businesses to find and present user-generated content to their target audiences on their own online marketing channels, as part of an overall content marketing strategy (Valos et al., 2010). The small business owner-managers in this study recognised the need for this longer term approach to online customer relationship building, referring to softer metrics that are needed to measure interactions (e.g. retweets and replies on Twitter, comments on Facebook, follows on Pinterest). This demonstrates a good understanding of the concept of the customer journey and concurs with Valos et al. (2010), who found that marketers are more accepting of soft attitudinal measures alongside harder quantitative measures (e.g. of sales).

Phase 3 is a test-and-learn phase, designed to encourage owner-managers to monitor closely the impact of their marketing and to refine their goals. This study has demonstrated that the ability to closely measure impact is central to owner-managers retaining a clear focus. It is also central to them taking more control of their future which is a concept tied to effectuation. Sarasvathy's (2001) research found that successful entrepreneurs created rather than predicted the future. Prediction, in this increasingly complex digital landscape, is challenging for large organisations and beyond the scope of smaller enterprises. Creating a test-and-learn culture among smaller enterprises offers the potential for them to identify new goals based on the extended means (who they know and what they know) they have at their disposal. This extended means is enabled by better knowledge of customer, the ability to create new customer-focused products and services, enabled where appropriate through value-added partnerships with other service providers. The findings reveal that owner-managers are more comfortable with shorter term planning cycles and this model is designed to enable more agile marketing (Breur, 2011; Chaffey, 2010). However, the lack of knowledge and, in some cases, confidence, of owner-managers is a barrier to achieving a higher level of agility and gives credence to the caution sounded by Valos et al. namely that marketers need more 'sophisticated campaign measurement systems' (2010, p. 369).

Consequently, all phases of the model require that owner-managers develop technical skills to support their ability to track and measure online customer engagement (Chaffey & Patron, 2012; Soonsawad, 2013).

Conclusions and future research

This study has explored the adoption of technology for marketing by 24 small business owner-managers using a qualitative methodology. The study has illustrated that, while the speed of technological innovations pose a knowledge gap for some owner-managers, they are aware of the importance of adopting technology for marketing and recognise the opportunities it affords. The owner-managers that participated in this study displayed a willingness to embrace technology and their current adoption of technology for marketing revealed interesting insights into the ways in which it could enable their businesses to be more competitive.

However, this predisposition towards technology was hindered and tempered by a lack of both technical and marketing competency, which, in today's digital environment, are increasingly interconnected. Our study, in common with Chaffey (2010), highlighted the difficulties of measuring ROI in marketing technology and the 'dearth of research or frameworks for analytical approaches to assess and plan investment in digital media' (Chaffey, 2010, p. 191).

This study has identified several potentially interesting questions for further research that would enrich our understanding of owner-manager technology adoption and help to improve practice in this area.

Firstly, how can effectuation, which most owner-managers tend towards, be enabled by the wealth of online marketing analytical data that is available? This of course is dependent on the technical ability of owner-managers to gather and use this data effectively, which brings into focus a second question: how might owner-managers best acquire those skillsets? A promising line of inquiry suggested by this study is to explore the role that small business networks could play in enabling owner-managers not only to acquire the knowledge in the short term but to sustain it in the longer term. This is particularly important given the speed of technological change. Thirdly, we have proposed a model based on the generalisations we can derive from the analysis of the 24 businesses in this study. Clearly, additional research is required to now further test the model and its wider applicability to other industry sectors outside the wider service economy represented in this study. Finally, and returning to effectuation, a further area for research would be to measure the impact that this model has on the performance of owner-managed enterprises in a number of areas, for example, business agility, innovation, entrepreneurship, and business growth.

Disclosure statement

No potential conflict of interest was reported by the authors.

Funding

This work was supported by the Economic & Social Research Council [grant number ES/J020990/1].

References

Ansari, A., & Mela, C. F. (2003). E-customization. *Journal of Marketing Research, 40*(2), 131–145.
Barnes, D., Clear, F., Dyerson, R., Harindranath, G., Harris, L., & Rae, A. (2012). Web 2.0 and micro-businesses: An exploratory investigation. *Journal of Small Business and Enterprise Development, 19*(4), 687–711.
Bazeley, P., & Jackson, K. (2013). *Qualitative data analysis with NVivo.* London: Sage.
de Berranger, P., Tucker, D., & Jones, L. (2001). Internet diffusion in creative micro-businesses: Identifying change agent characteristics as critical success factors. *Journal of Organizational Computing & Electronic Commerce, 11*(3), 197–214.

Bharadwaj, P., & Soni, R. (2007). E-commerce usage and perception of e-commerce issues among small firms: Results and implications from an empirical study. *Journal of Small Business Management, 45*(4), 501–521.

Bharati, P., & Chaudhury, A. (2006). Studying the current status: Examining the extent and nature of adoption of technologies by micro, small and medium sized manufacturing firms in the greater Boston area. *Communications of the ACM, 49*(10), 88–93.

Black, H., & Kelley, S. (2009). A storytelling perspective on online customer reviews reporting service failure and recovery. *Journal of Travel & Tourism Marketing, 26*(2), 169–179.

Breur, T. (2011). Data analysis across various media: Data fusion, direct marketing, clickstream data and social media. *Journal of Direct, Data and Digital Marketing Practice, 13*(2), 95–105.

Bulearca, M., & Bulearca, S. (2010). Twitter: A viable marketing tool for SMEs? *Global Business & Management Research, 2*(4), 296–309.

Chaffey, D. (2010). Applying organisational capability models to assess the maturity of digital-marketing governance. *Journal of Marketing Management, 26*(3/4), 187–196.

Chaffey, D., & Ellis-Chadwick, F. (2012). *Digital marketing strategy, implementation and practice.* Harlow: Pearson Education.

Chaffey, D., & Patron, M. (2012). From web analytics to digital marketing optimization: Increasing the commercial value of digital analytics. *Journal of Direct, Data and Digital Marketing Practice, 14*(1), 30–45.

Chaston, I., Badger, B., Mangles, T., & Sadler-Smith, E. (2003). Relationship marketing, knowledge management systems and e-commerce operations in small UK accountancy practices. *Journal of Marketing Management, 19*(1/2), 109–129.

Coviello, N., Milley, R., & Marcolin, B. (2001). Understanding IT-enabled interactivity in contemporary marketing. *Journal of Interactive Marketing, 15*(4), 18–33.

Dandridge, T., & Levenburg, N. M. (2000). High-tech potential? An exploratory study of very small firms' usage of the Internet. *International Small Business Journal, 18*, 81–91.

Dhar, V., & Chang, E. (2009). Does chatter matter? The impact of user-generated content on music sales. *Journal of Interactive Marketing, 23*(4), 300–307.

Durkin, M., McGowan, P., & McKeown, N. (2013). Exploring social media adoption in small to medium-sized enterprises in Ireland. *Journal of Small Business & Enterprise Development, 20*(4), 716–734.

Fillis, I., & Wagner, B. (2005). E-business development: An exploratory investigation of the small firm. *Die Entwicklung des E-Business: Eine explorative Untersuchung des Klein betriebs, 23*(6), 604–634.

Fischer, E., & Reuber, R. (2011). Social interaction via new social media: (How) can interactions on Twitter affect effectual thinking and behavior? *Journal of Business Venturing, 26*, 1–18.

Harrigan, P., Schroeder, A., Qureshi, I., Yulin, F., Ibbotson, P., Ramsey, E., & Meister, D. (2010). Internet technologies, ECRM capabilities, and performance benefits for SMEs: An exploratory study. *International Journal of Electronic Commerce, 15*(2), 7–46.

Harris, L., & Rae, A. (2009). Social networks: The future of marketing for small business. *Journal of Business Strategy, 30*(5), 24–31.

Hinchcliffe, D. (2010). Why all the fuss about Web 2.0? *Infonomics, 24*(1), 26–31.

Hisanabe, Y. (2009). Persona marketing for Fujitsu kids site. *Fujitsu Scientific and Technical Journal, 45*(2), 210–218.

Jones, C., Hecker, R., & Holland, P. (2003). Small firm Internet adoption: Opportunities forgone, a journey not begun. *Journal of Small Business and Enterprise Development, 10*(3), 287–298.

Jones, P., Simmons, G., Packham, G., Beynon-Davies, P., & Pickernell, D. (2014). An exploration of the attitudes and strategic responses of sole proprietor micro-enterprises in adopting information and communication technology. *International Small Business Journal, 32*, 285–306.

Kara, A., Spillan, J., & DeShields, O. (2005). The effect of a market orientation on business performance: A study of small-sized service retailers using MARKOR scale. *Journal of Small Business Management, 43*(2), 105–118.

Kim, H., Lee, I., & Lee, C. (2011). Building Web 2.0 enterprises: A study of small and medium enterprises in the United States. *International Small Business Journal, 31*(2), 156–174.

Malthouse, E., Haenlein, M., Skiera, B., Wege, E., & Zhang, M. (2013). Managing customer relationships in the social media era: Introducing the social CRM house. *Journal of Interactive Marketing, 27*(4), 270–280.

Martin, L., & Matlay, H. (2003). Innovative use of the Internet in established small firms: The impact of knowledge management and organisational learning in accessing new opportunities. *Qualitative Market Research*, *6*(1), 18–26.

McGowan, P., & Durkin, M. (2002). Toward an understanding of Internet adoption at the marketing/ entrepreneurship interface. *Journal of Marketing Management*, *18*(3/4), 361–377.

O'Reilly, T. (2007). What is Web 2.0: Design patterns and business models for the next generation of software. *Communications & Strategies*, 1st Quarter, (65), 17–37.

Payne, A., & Frow, P. (2005). A strategic framework for customer relationship management. *Journal of Marketing*, *69*, 167–176.

Pelham, A., & Wilson, D. (1996). A longitudinal study of the impact of market structure, firm structure, strategy, and market orientation culture on dimensions of small-firm performance. *Journal of the Academy of Marketing Science*, *24*(1), 27–43.

Peltier, J., Zhao, Y., & Schibrowsky, J. (2012). Technology adoption by small businesses: An exploratory study of the interrelationships of owner and environmental factors. *International Small Business Journal*, *30*(4), 406–431.

Polo Peña, A., Frías Jamilena, D., & Rodriquez Molina, M. (2011). Impact of market orientation and ICT on the performance of rural smaller service enterprises. *Journal of Small Business Management*, *49*(3), 331–360.

Qualman, E. (2013). *Socialnomics*. Hoboken, NJ: Wiley.

Read, S., Sarasvathy, S. D., Dew, N., Wiltbank, R., & Ohisson, A. (2011). *Effectual entrepreneurship*. London: Routledge.

Rhodes, J. (2009). A strategic framework for rural micro-enterprise development: The integration of information communication technology (ICT), e-commerce, marketing, and actor-network theory. *Perspectives on Global Development & Technology*, *8*(1), 48–69.

Ritchie, B., & Brindley, C. (2005). ICT adoption by SMEs: Implications for relationships and management. *New Technology, Work & Employment*, *20*(3), 205–217.

Sarasvathy, S. (2001). Causation and effectuation: Toward a theoretical shift from economic inevitability to entrepreneurial contingency. *The Academy of Management Review*, *26*(2), 243–263.

Sellitto, C., Andrew, W., & Burgess, S. (2003). A review of the web sites of small Australian wineries: Motivations, goals and success. *Information Technology and Management*, *4*(2/3), 215–232.

Sigala, M. (2006). e-Customer relationship management in the hotel sector: Guests' perceptions of perceived e-service quality levels. *Tourism*, *54*(4), 333–344.

Simmons, G., Armstrong, G., & Durkin, M. (2008). A conceptualization of the determinants of small business website adoption: Setting the research agenda. *International Small Business Journal*, *26*(3), 351–389.

Simmons, G., Armstrong, G., & Durkin, M. (2011). An exploration of small business website optimization: Enablers, influencers and an assessment approach. *International Small Business Journal*, *29*(5), 534–561.

Soonsawad, P. (2013). Developing a new model for conversion rate optimization: A case study. *International Journal of Business and Management*, *8*(1), 41–51.

Standing, C., Vasudavan, T., & Borberly, S. (1998). Re-engineering travel agencies with the world wide web. *Electronic Markets*, *8*(4), 40–43.

Street, C., & Cameron, A. (2007). External relationships and the small business: A review of small business alliance and network research. *Journal of Small Business Management*, *45*(2), 239–266.

Tussyadiah, I., & Fesenmaier, D. (2008). Marketing places through firstperson stories – an analysis of Pennsylvania roadtripper blog. *Journal of Travel & Tourism Marketing*, *25*(3/4), 299–311.

Valos, M., Ewing, M., & Powell, I. (2010). Practitioner prognostications on the future of online marketing. *Journal of Marketing Management*, *26*(3/4), 361–376.

Verreynne, M., Parker, P., & Wilson, M. (2013). Employment systems in small firms: A multilevel analysis. *International Small Business Journal*, *31*(4), 405–431.

Wolcott, P., Kamal, M., & Qureshi, S. (2008). Meeting the challenges of ICT adoption by micro-enterprises. *Journal of Enterprise Information Management*, *21*(6), 616–632.

Wymer, S., & Regan, E. (2005). Factors influencing e-commerce adoption and use by small and medium businesses. *Electronic Markets*, *15*(4), 438–453.

Exploring interactive communication using social media

Chrystal B. Zhang[a] and Yi Hsin Lin[b]

[a]*Faculty of Science, Engineering & Technology, Swinburne University of Technology, Melbourne, Australia;* [b]*Department of Leisure & Recreation Management, Asia University, Wufeng, Taiwan*

Engaging consumers in interactive marketing communication is instrumental in business–customer relationships building and development. Social media enables consumers to initiate marketing messages and gain growing control in the communication process due to its enhanced interactivity features. This study investigates whether communication between businesses and their social media users is interactive and how interactive it is. It also attempts to determine what types of messages are more likely to result in interactive communication. The findings reveal that businesses are attempting functional interactivity while individual users are increasingly securing a control to achieve contingent interactivity. Businesses are adopting a consumer-centric approach in designing and executing marketing communication messages to achieve interactivity though significant variance is established between businesses as measured by Interactivity Performance Matrix. Challenges remain for businesses as to how to engage their customers by utilizing the interactivity features of social media to facilitate relationships formulation and development.

Introduction

Marketing communication has experienced a paradigm shift over the last two decades from integrated marketing communication focusing on businesses that drive the integration and conveyance of a consistent message to consumers to relationship communication which advocates a consumer-centric approach (Finne & Gronroos, 2009; Vlasic & Kesic, 2007). While businesses are only able to create the circumstances, through a comprehensive planning, for their marketing communication message to be conveyed (Finne & Gronroos, 2009), it is consumers who hold the key to enabling the integration of such communication, hence making the communication relational resulting in relationship formulation and development. Non-interactive marketing communication, without the involvement and participation of consumers, is set to evolve into interactive communication with a more conversational style of two-way dialogue between businesses and consumers (Deighton & Grayson, 1995).

Technology advancement, such as the World Wide Web, has enabled consumers to take part in the marketing communication process, resulting in interactive conversational communication (Pantelidis, 2010; Schmallegger & Carson, 2008). It has opened up a new door for businesses to engage with their customers in new and meaningful ways (Durkin, Filbey, & McCartan-Quinn, 2014) and to maintain an instant and constant dialogue at lower cost, thus facilitating the achievement of relationship communication between businesses and consumers.

The latest development of such technology advancement is social media. Social media is defined as a group of Internet-based applications (Kaplan & Haenlein, 2010) and comprises both the conduits that can carry the content and the content itself that can be presented in various formats (Berthon, Pitt, Plangger, & Shapiro, 2012). Regardless of being a business or a consumer, social media allows every user to enjoy the equal opportunity to create, contribute and disseminate messages (Li & Wang, 2011; Thevenot, 2007). Such user-created content has unprecedentedly enriched communication, both in its content and in format.

In comparison to static websites, social media has enabled real-time conversations between all users without time or physical constraints. This promises an increased online interactive communication, allows a relationship to be developed and sustained (Etter & Fieseler, 2010), generates sales and revenue (Kunz & Hackworth, 2011), and aids in strengthening customer loyalty and improving the overall satisfaction (Kasavana, Nusair, & Teodosic, 2010).

Envisaging the greatest potential, businesses are quick to capitalize on the properties of social media by exploiting its prospective multifaceted functions. For instance, some businesses have utilized social media as an advertising medium to promote products and services to a wider customer base (Hansson, Wrangmo, & Søilen, 2013). Others have attempted to use social media as a communication channel to provide customer support, keep their customers updated about their products and services and publish news releases, thus staying connected with their customers to create and strengthen the bond between them (Mitic & Kapoulas, 2012).

While social media has presented enormous opportunities for businesses in a dynamic way, it also posits unprecedented challenges. One of the key concerns of businesses is that they are deprived of the privilege to control the communication process with their well-designed public relation messages, as social media bestows the once passive customers the equal power to participate in and contribute to the communication. Another concern is the effect of 'viral marketing' or 'social media marketing' (Kozinets, de Valck, Wojnicki, & Wilner, 2010), as used interchangeably, that allows a message to be transmitted to a large group of people instantly (Kaplan & Haenlein, 2011), which could be either instrumental or detrimental to a business brand. Furthermore, businesses have not fully understood what constitutes the multifaceted concept of interactivity in the virtual environment (Johnson, Bruner, & Kumar, 2006) and the shortage of resources to manage the online activities has hindered any strategies to be formulated and any proactive actions to be taken effectively. The lack of interaction between businesses and their customers on social media could affect the ultimate success of corporate strategies and the business competitive edge (Chan & Denizci Guillet, 2011).

To date, a burgeoning mountain of literature on social media has emerged, examining how businesses have exploited its wide array of functions. However, there is a gap in the available literature relative to the efficacy of using social media as a communication mechanism to facilitate interactive relationship communication between businesses and their customers. Specifically, the majority of social media literature to date still takes a business-centric approach (sender) based on an inside-out view by concentrating their analyses on business' strategies and actions of employing social media in their marketing activities. Businesses are still considered to control the craft of marketing communication while consumers are treated as passive recipients, though creative and influential.

While there is some scholarship examining consumers' behaviour on social media, the focus is on their motives, personality traits and utilization patterns in using social media. A handful of studies have attempted to investigate the contribution of consumers to social media communication, but the findings are more or less empirical evidence of the types of messages that are created and shared on social media and its impact on the corporate brand. An outside-in approach focusing on consumers is yet to be adopted, and consumers' role in marketing communication has not been fully recognized to the same extent as that of the businesses (Finne & Gronroos, 2009). As a consequence, few studies are able to investigate one step further to establish whether it is both businesses and consumers who equally contribute to the marketing communication on social media resulting in interactive communication, how interactive such communication is, and whether such interactive communication contributes to business–customer relationships building and development.

The lack of any informative and conclusive results in this aspect is likely to be due to the following. Firstly, there is an ongoing debate as to how to define the interactive communication between businesses and their customers in an online environment. Researchers have thus to determine and interpret the term in the context of the subject matter in question, resulting in a diversity of definitions which cannot be universally accepted (Blasco-Arcas, Hernandez-Ortega, & Jimenez-Martinez, 2014). Next, there is a variance of measurement when determining the interactive communication between businesses and their customers. Some propose the perceived technology-enabled people–machine interactivity while others contest to examine interactivity between users by treating technology as an enabler. In addition, the role of interactive communication in customer relationship development and management remains an intriguing interdisciplinary area stretching across communication and marketing, which therefore requires more empirical research in order to validate propositions and facilitate theory building and development.

While it is not disputed that interactive communication plays a pivotal role in developing and maintaining business–customer relationships, a consumer-centric approach to marketing communication should be advocated to ensure that consumers create the same meaning as intended by the businesses to enable the successful integration of marketing messages (Finne & Gronroos, 2009) and relationships formulation. Technology advancement has offered customers powerful avenues through which to participate and play an active role in marketing communication in online environment (Blasco-Arcas et al., 2014). There is a compelling need to examine the marketing communication from both businesses and customers' perspectives, who are cooperating as well as competing in contributing to the interactive communication. This study therefore proposes to investigate the communication on social media without discriminating consumers' contribution. We aim to establish whether both businesses and their customers contribute to marketing communication on social media and further determine whether, and to what extent, such communication is interactive and conducive to relationship building and development. We use the Facebook postings collected from 20 airlines in five regions as the object of the study. Specifically, we attempt to answer the following research questions (RQs).

RQ1. What types of messages are generated by, and exchanged between, businesses and their customers on Facebook?

RQ2. Are the messages interactive?

RQ3. How interactive is the communication?

RQ4. What types of messages are more likely to result in interactive communication leading to relationships formulation and development?

Relationship marketing, interactive communication and social media

Marketing has become more and more communication dependent in that communication has become the primary integrative element in managing a business–customer relationship (Duncan & Moriarty, 1998; Kelleher, 2009; Ledingham, 2003). Relationship marketing embraces a collection of all marketing activities directed towards establishing, developing and maintaining successful relational exchanges (Morgan & Hunt, 1994; Yoon, Choi, & Sohn, 2008). As a theoretical concept and industry practice, it has thus gained its threshold over the last two decades and has since received close attention due to its importance of maintaining and enhancing existing and prospective business–customer relationships (Duncan & Moriarty, 1998). Two-way communication and interactivity is instrumental in this relationship-building process. It has demonstrated capability of engaging both businesses and their customers, enabling customers to create shared meaning of marketing messages as intended by businesses, and allowing the 'new generation' of derivative marketing approaches, such as customer-focused, market-driven, outside-in, data-driven marketing, integrated marketing and integrated marketing communications, to emerge and develop (Cross & Smith, 1995; Duncan & Moriarty, 1998).

Marketing communication becomes interactive when all parties concerned are engaged with each other through participating in all types of activities that would convert the marketing communication messages and create meaning in a mutually beneficial way that affects the knowledge base between the parties (Lindberg-Repo, 2001). An interactive communication is found to be positively associated with the perceived relationship investment of the businesses in terms of their efforts, resources and attention aimed at maintaining or enhancing relationships with the customers (Yoon et al., 2008). Interactive communication can occur both in technologically mediated (e.g. the Internet) and unmediated (e.g. face-to-face) contexts. Media, whose communication forms come closest to face-to-face, are traditionally considered the most interactive. Heeter (1989) asserts that the more the technologically mediated communication resembles interpersonal communication, the more interactive the communication is. However, Gulbrandsen and Just (2011) argue that technology does not automatically generate interactive communication. Rather it is the ability and willingness of the users to participate and contribute that enables an interactive communication. Although various interactive features are equally available on a website, consumers tend to take advantage of those features in different ways and their practices of using the features do change in relation to contextual matters (Boczkowski & Mitchelstein, 2012). Businesses, therefore, need to pay attention to the media features, users' practices and their interdependences in order to achieve interactivity and improve its outcome (Boczkowski & Mitchelstein, 2012).

Sundar, Kalyanaraman, and Brown (2003) suggest two types of technology-mediated interactivity: functional interactivity and contingency interactivity. A functional interactivity is basically an interface's capacity of conducting a dialogue or information exchange between users and the interface. Features such as the presence of an e-mail link, chat room, event calendar, search function, and functions hosting surveys, polls and any other format inviting users to send responses to the web-hosting organization are said to be essential in facilitating a dialogue loop, which in turn encourages visitors to action when visiting or returning to the web, thus resulting in interactivity. A contingency interactivity, however, is a process involving users, media and messages in which communication roles are interchangeable for full interactivity to occur. Contingency means that messages in an interactive process of communication are contingent upon previous messages. The more a user's response to another and the more intertwined and cumulative, the more interactive the communication is (Rafaeli, 1988; Walther, Gay, & Hancock, 2005).

Social media is the latest development of Internet-facilitated communication platform that has the potential to enhance interactive communication to a higher level. In the traditional, offline environment, marketing communication is initiated by businesses and relationships between businesses and their customers are manoeuvred through various promotional efforts such as advertising, direct mail and sales promotions (Yoon et al., 2008). The Internet, however, empowers the consumers with the initiative to establish and retain relationships with businesses by proactively engaging in two-way, self-driven communication and exercising an independent control over the communication process (Bezjian-Avery, Calder, & Iacobucci, 1998; Yoon et al., 2008). Both businesses and consumers have control and may in turn act as senders and receivers (Fred Van Raaij, 1998). Social media has extended the capacity of the Internet by enabling instant and constant, unique and personalized conversations between businesses and their customers, thus making the interactive communication more achievable and effective. Kelleher (2009) asserts that interactive and participatory features of social media facilitate businesses to realize the potential of the Internet to achieve relational outcomes, which include trust, satisfaction, community and control mutuality.

While few will question the interactive feature of social media, especially its promising role of enhancing interactive communication to a higher level, little is known of whether businesses and their customers have explored this interactivity feature and how they are engaging with each other resulting in interactive communication. Swain (2005) points out that industry practitioners and academics alike are, to some extent, sceptical of the dominance of interactivity in marketing communication though they anticipate that its importance will grow and it will take some time for such activities to be extensively adopted and incorporated into practitioners' daily responsibilities. Burton and Soboleva (2011) have established that some companies have just started to engage with customers in an interactive communication, though there is a large variance observed across the industries. They note that businesses tend to struggle when handling conversations and responses to their customers due to their unclear social media strategies, lack of internal coordination, and short of resources and skills required (Burton & Soboleva, 2011). Etter (2013) discovers that businesses are hesitant to engage into an interaction with their customers on Twitter when the topic is more sensitive, is a contested area or has a major impact on its reputation and legitimacy. They therefore tend to adopt a cautious approach to interactive communication to mitigate any possible criticism or risks of attracting critical stakeholders that would openly question its legitimacy (Etter, 2013).

Baird and Parasnis (2011) argue that there exists a misperception in the business community about users' willingness to communicate and connect with brands on Facebook. The majority of consumers treat Facebook as a vehicle for personal connections with friends and families rather than engaging with businesses (Baird & Parasnis, 2011). They are less interested in the organizations on Facebook, nor are they enthusiastic about communicating with organizations (Vorvoreanu, 2009). While consumers' key motives in choosing to use Facebook are entertainment, passing time, self-expression, information seeking and gaining self-satisfaction (Hunt, Atkin, & Krishnan, 2012; Leung & Bai, 2013; Zywica & Danowski, 2008), their capacity of using social media is dictated by their personality traits, gender differences and interpersonal competency (Hunt et al., 2012; Jenkins-Guarnieri, Wright, & Hudiburgh, 2012).

Swain (2005) argues that it is reasonable to speculate that the consumer-initiated, controlled and maintained interactive communication with businesses will escalate eventually

(Swain, 2005) as consumers have demonstrated a positive attitude towards interactive marketing communication (Vlasic & Kesic, 2007). Consumers perceive more added value to themselves in a more personalized interactive communication but more value to businesses in an automated interactive communication. The challenge is whether the consumers perceive any added value to compensate for the possible costs incurred, as well as for the extra efforts invested in mastering and using the interactive technologies. Consumers are only prepared to accept any new types of communications with businesses if they see any adequate added value (Vlasic & Kesic, 2007).

When using social media to convey marketing messages, businesses are found to treat social media in a more conventional marketing approach. One such example is that the language used for social media advertisements and sales promotions is very one-way static statements, with the overall tone leaning toward formal, official and authoritative (Hvass & Munar, 2012). The information is 'broadcasted' rather than conversed (Carim & Warwick, 2013), with customers being less inspired to participate. Such an authoritative tone adopted by businesses demonstrates that a command-and-control attitude towards social media is still prevailing, which would limit the interactivity between posters, hence hindering relationships building and development (Shih, 2009). Merchant, Elmer, and Lurie (2011) argue that social media is more than just another mass or interpersonal communication channel such as print, television or radio. Rather, it is the polar opposite of traditional marketing media. While the added value of social media is its capacity to enable participatory and interactive communication, businesses need to reconsider how to develop and disseminate their marketing messages on social media. Schmallegger and Carson (2008) rightly suggest that business-created content on social media should not be promotion-oriented. Instead, it ought to provide some added value to customers (Akehurst, 2009; Schmallegger & Carson, 2008) so as to be more inviting and engaging.

Methodology

Social media site selection

Of all the social media sites, Facebook is the most popular and takes the biggest market share. It embraces the interactivity of the Internet by allowing users to utilize various features designed for interpersonal communication (Hunt et al., 2012) appealing to both businesses and individual users. Specifically, it acts as an interface allowing businesses to post any messages attracting users to visit the site and take further action by clicking, emailing, sharing and commenting, thus achieving a functional interactivity. It also serves as a contingency interactivity platform to enable two-way communication that can develop into a thread of discussion between businesses and their users, and between users themselves.

Airlines selection

Two full service carriers (FSCs) and two low cost carriers (LCCs) were chosen from five regions, bringing the total sample to 20. In determining which FSC and LCC to select, two criteria applied. The first was the winner of the top Skytrax World Airline Awards 2011 of the category for the region, and the second was the number of 'likes' the airline's Facebook page has received from their Facebook users. The commodity they share is that they are the most liked carriers by passengers who express their views in different ways. The Skytrax Airline Award is voted by travellers from over 160 countries in airline passenger satisfaction surveys and considered the most prestigious global benchmark for excellence of service. The Facebook 'likes' is uniquely designed to allow individual users to express their appreciation or a positive view on the company's products and services without necessitating a written comment. The more 'likes' a company receives on its Facebook page, the more positive the company's image is perceived though not necessarily a reflection of the best products and services provided. Table 1 is the list of airlines selected for this study.

Posts collection

Posts were collected between 16th and 23rd August 2012, from the English version of the Facebook page of the above 20 airlines. For example, KLM Royal Dutch Airlines' (KLM) posts were collected from the English version of its Facebook page in the Netherlands rather than the Dutch version. The timing chosen is believed to be appropriate in that it was a typical operational week without reported accidents, incidents or crises on public domains. The posts were not closely related to any specific issues or crises that were pertinent to any particular airline, hence were an authentic collection of the messages authored by both airlines and their Facebook users.

Coding and categorizing themes of postings

We first grouped all posts into two broad categories: airline-authored content and individual user-authored content. We then divided airline-authored content into two subgroups: airline-initiated and airline-responded posts. Airline-initiated posts are messages initiated by airlines, which contain the information that airlines want to communicate to their Facebook users. Airline-responded posts are messages responded by airlines to their users, though airlines retained authorship of such messages.

We then analyzed each group of posts in an attempt to uncover a theme pertinent to the messages. Initially, the themes developed by Sreenivasan, Lee, and Goh (2012) were applied to our collection, with reference being made to readings in marketing literature as well as definitions and scope adopted by Hvass and Munar (2012). However, the groupings were modified for a more accurate and appropriate reflection of the messages. Headings bearing identical themes were integrated into one category, and new entries were added to represent novel themes. It is believed that such regroupings are both essential

Table 1. List of 20 airlines selected for the study.

Criteria	Airlines		Europe	Oceania	North America	Africa	Asia/Middle East
Skytrax Award 2011 winner of the region	FSCs	Swiss International Air Lines	Air New Zealand	Air Canada	South Africa Airways	Qatar Airways
	LCCs	easyJet	Jetstar Airways	Virgin America	Kulula Airlines*	AirAsia
Facebook likes	FSCs	KLM	Qantas Airways	United Airlines	EGYPTAIR	Jet Airways
	LCCs	Ryanair	Tigerair Australia	Southwest Airlines	Mango Airlines*	Tigerair Singapore

*Refers to the airlines that were chosen based on Facebook likes alone as there is no award available in the region.

and appropriate since both deductive and inductive methods were employed to ensure the fine-tuned development and accurate reflection of all themes, which in turn warrants an objective, consistent and systematic analysis of the content. A modified list of seven categories of themes was eventually suggested for both airline-initiated and user-authored posts. Table 2 is a detailed description of the categories of the themes of the posts and their definitions. Coding was conducted by a group of five undergraduate students at Swinburne University of Technology with the guidance and support of the first author of this paper. Where disagreement arose, the team discussed and referred to the definitions for an agreement.

Determining whether the communication is interactive or non-interactive

In determining whether the communication is interactive or non-interactive, we applied the definitions and methods discussed in the literature review, whereby interactivity includes both functional and contingent interactive communication. When a post was a poll, a quiz, a contest or a call for participation in a vote, it was classified as functional interactivity. When a post led to a thread of discussion and message exchanges between airlines and their users, it was classified as contingent interactivity. If a post did not bear any of the above features, it was treated as non-interactive communication.

Table 2. Categories of airline-initiated and user-authored posts and their respective definition.

Airline-initiated post		User-authored post	
Category	Definition	Category	Definition
Advertising, sales and promotion	Promotion of airlines products or services, with or without a link to purchase	Compliment	Acknowledgement and appreciation of airlines service
Information provision	Response to users' queries and questions, provision of clarifications	Grievance	Negative feedback, complaints about airlines product and services
Sponsorship	Information about an airline-sponsored event or initiative	Sharing travel experience	Sharing user's experience with stories or photographs, without feedback received from community
Real-time updates	Timely updates on airlines' products, services, facilities and other information that is relevant to the users' travel and trips	Information seeking	Raising a query and expecting an response and clarification
Social activity	Initiating a dialogue without specific information about the company	Community support	Sharing information and/or answering queries raised within the user community
Poll/contest	Invitation to users to participate a quiz or bid for a prize by answering a test	Social activity	Personal updates, airline–users interactions and users community interaction
Miscellaneous	Posts that do not fall into any of the above categories	Miscellaneous	Posts that do not fall into any of the above categories

To measure the performance of interactive communication of the sampled airlines, we developed an Interactivity Performance Matrix. The vertical axis refers to the number of airline-authored interactive messages while the horizontal axis represents the number of user-authored messages. We name the quadrants as Leader at the upper right, User-activist at the lower right, Business-activist at the upper left and Laggard at the lower left. Leader is a carrier that is interactively communicating (both functional and contingent) with its users, and its users are also enthusiastically and willingly posting on the carrier's Facebook page. User-activist is an airline whose users are more active in contributing to the page but the carrier is less responding. Business-activist is a carrier that is more willing to engage with its users but, unfortunately, its users are not motivated or less interested in being involved. Laggard is the one which neither the business itself nor its users are active in engaging with each other on Facebook.

Results

During the data collection period, airlines authored 1875 posts in comparison to users' 2990 posts. Table 3 is a breakdown of the posts created by both airlines and their respective users. Among all the airlines, KLM authored more than one-third (35.9%) of the posts followed by Jetstar Airways (11.5%). Ryanair was the least active airline generating zero posts while Tigerair Singapore was the second least active carrier contributing a meagre 0.9% of posts. From the users' perspective, KLM's users contributed the most, almost a quarter (19.6%) of posts, followed by Southwest Airlines (15.7%). Three airlines, Jet Airways, Qatar Airways and Tigerair Singapore, had no user-created posts recorded.

Table 3. Breakdown of airline-authored and user-authored posts.

Airlines	Airline-authored				User-authored	Percentage
	Airline-initiated	Airline-responded	Total	Percentage		
KLM	9	663	672	35.9	587	19.6
Swiss International Air Lines	11	27	38	2.0	46	1.5
United Airlines	2	27	29	1.6	323	10.8
Air Canada	4	187	191	10.2	203	6.8
South Africa Airways	4	4	8	0.4	106	3.6
EGYPTAIR	31	94	125	6.7	55	1.8
Jet Airways	13	8	21	1.1	0	0.0
Qatar Airways	12	2	14	0.8	0	0.0
Qantas Airways	8	77	85	4.5	196	6.6
Air New Zealand	10	16	26	1.4	72	2.4
Ryanair	0	0	0	0.0	66	2.2
easyJet	2	80	82	4.4	142	4.8
Virgin America	8	46	54	2.9	57	1.9
Southwest Airlines	16	32	48	2.6	469	15.7
Kulula Airlines	3	68	71	3.8	59	2.0
Mango Airlines	17	65	82	4.4	46	1.5
Tigerair Singapore	3	0	3	0.2	0	0.0
AirAsia	11	83	94	5.0	219	7.3
Jetstar Airways	3	212	215	11.5	308	10.3
Tigerair Australia	5	12	17	0.9	36	1.2
Total	172	1703	1875	100.0	2990	100.0

Categories of themes of airline-initiated messages

Of the 1875 airline-authored posts, 172 were airline-initiated posts and 1703 were airline-responded posts. FSCs contributed 60% of airline-initiated posts with the other 40% coming from the LCCs. Figure 1 is the comparison of categories of themes of posts initiated by FSCs and LCCs. Of these 172 posts, 45.3% fell into the category of 'Advertising, and sales and promotion', for example, statements created by airlines announcing sales and promotions of their existing destinations. LCCs were more active in advertising its services on Facebook, with around 57.4% of all of its posts contributing to this category. This finding echoes the observation made by other researchers such as Kunz and Hackworth (2011) and Hays, Page, and Buhalis (2013) that advertising is the key activity undertaken by businesses on social media.

Some messages included a hyperlink that could take Facebook users directly to a flight booking page by a simple click. For example, United Airlines posted 'People wear "1 < 3 NY" shirts for a reason. What is yours? Experience summer in the Big Apple with our New York fare. http://bit.ly/O9m6tN'. Others specified conditions for taking advantage of the special promotions without a hyperlink. For example, Tigerair promoted its routes between Singapore and Thailand for $47.00. 'Strip member's exclusive access: 6 August, 8am to 2pm. Public access: 6 August, 2.01pm to 8 August, 11.59pm, conditions apply'. This finding, though contradictory to what Hvass and Munar (2012) noted, demonstrates that airlines are making efforts to explore the distribution function of social media. Such a move bears strategic significance as distribution remains one of the biggest cost sources for airlines. Should social media have the potential and capacity to improve efficiency resulting in cost reductions, any efforts are worthwhile.

'Social activity' was the second largest category accounting for 27.9%, with FSCs demonstrating far more enthusiasm for this category. These posts, though not necessarily directly related to the provision of air services, aimed at promoting goodwill and publicity. The content ranged from acknowledging local cultural holidays to celebrating sports events, such as the 2012 London Olympics. Exemplars include Qantas Airways that posted an album of 21 photos featuring Australia's Olympic athletes boarding its flights back home and South African Airways who boasted of their pride in flying their Olympians to London and back home. 'Poll/contest' ranked third, taking 9.9% of the total airline-initiated

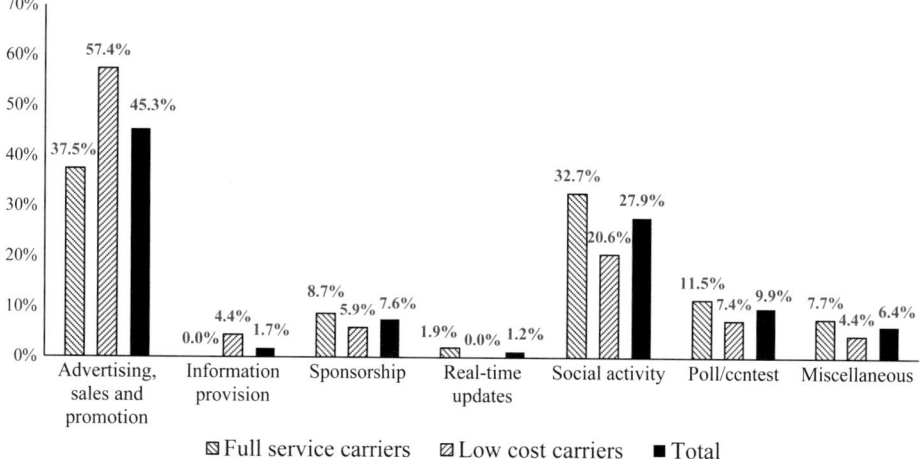

Figure 1. Comparison of categories of themes of posts initiated by FSCs and LCCs.

posts. This was an emerging theme which has been rarely documented in the existing literature. Sreenivasan et al. (2012) noted that about 8.5% of the tweets were polls and contests but they failed to give further discussion in their analyses. An airline's quiz on Facebook tended to relate to its product, aircraft used or destinations served. An incentive, such as a prize, was offered to motivate participation. The most appropriate answer to the quiz was given the following day.

The fourth category was 'Sponsorship' (7.6%) followed by 'Miscellaneous' (6.4%) and 'Information provision' (1.7%). The least active category was 'Real-time updates', which only took 1.2%, all contributed by FSCs. Despite this observation being inconsistent with the findings of Hvass and Munar (2012) that social media was used to disseminate time-sensitive information, it did not lead the researchers of this study to jump to the conclusion that airlines were not using social media to update their passengers about their services. A follow-up monitoring of KLM's and Southwest Airlines' respective Facebook pages revealed that both carriers did keep their passengers updated of their flight changes in the event of special circumstances. KLM informed its passengers of a system interruption that would prevent them from logging on to its website for online check-in and other services. Southwest Airlines advised its passengers of the prospective disruption to its services resulting from Hurricane Isaac which severely hit the southern part of the USA. The fact that few updates were collected from the sampled airlines in this study could be argued that during the data collection period, these airlines operated normally and routinely without irregular delays or interruptions, or with no introduction of any new procedures and policies. Hence, no update was required.

Categories of themes of messages authored by users

During the data collection period, users generously contributed 2990 posts. As shown in Figure 2, 'Information seeking' was the most popular category gaining 32.2% of all posts. Users have raised a wide array of queries which were either generic or specific. Generic queries sought general advice with the possibility of such advice given by both peer users and the airline concerned. For example, a user requested advice as to how to upsize their baggage limit. Another user asked for information on prospective sales and cheap fares from an airline. Specific queries usually related to a specific case, which necessitated a reply from the carrier in question. When asking specific queries, users

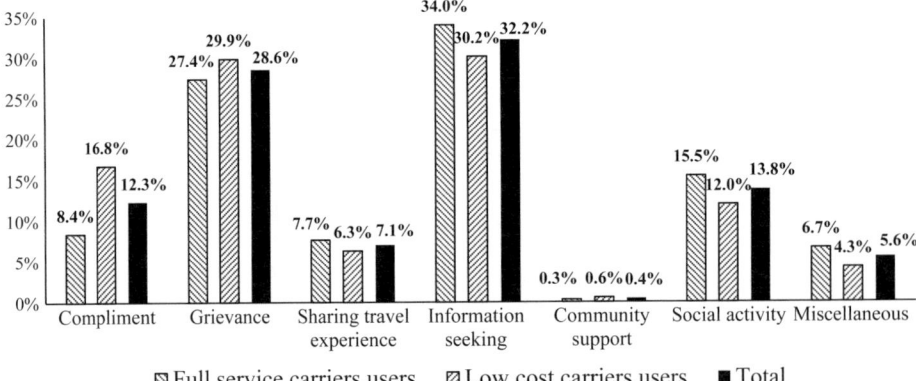

Figure 2. Comparison of categories of themes of posts authored by full service carrier users and low cost carrier users.

were quite assertive requesting the airline concerned to attend, address and update them of the progress of the outstanding case. For example, one user posted a query with a specific booking reference U92PUJ, asking AirAsia to provide an update as to how the case had been addressed. This finding echoes researchers such as Jansen, Zhang, Sobel, and Chowdury (2009) and Xiang and Gretzel (2010) who identified that social media is mainly used for information seeking.

'Grievance' was the second largest category of user-created posts accounting for 28.6%, while 'Compliment' ranked fourth, responsible for 12.3%. This finding contested Sreenivasan et al. (2012) who observed that users mainly used microblogs to render compliments rather than grievances. Although Facebook is a social networking site rather than a microblog and the data collection time is different, one thing was certain: users found social media sites the most convenient and effective platform to express themselves (Goh & Lee, 2011; Lerner, Han, & Keltner, 2007).

Of the 20 airlines monitored, 17 received more grievances than compliments, except KLM, Swiss International Air Lines and Southwest Airlines. Some grievances were a lengthy and detailed description of an awful personal travel experience. For example, one female user travelling with her daughter and granddaughter described how United Airlines failed to seat them together. Others could be negative comments, complaints or personal views on an airline's conduct as a response to other users' grievances. Such messages usually included a hyperlink to another story or description uploaded by other users. One such example was a story concerning United Airlines, which failed to look after an unaccompanied minor who did not arrive at the intended destination. This type of post is the most likely to become viral. Before social media was available, grievances and compliments from passengers were only heard when a survey was conducted or when users took the initiative to contact airlines via email or telephone. Often, the email account was generic, with no specific staff to address and the telephone number was most of the time engaged or there was no answer. Passengers become frustrated and discouraged in this process, patiently waiting for days before any response or even receiving no response from the airline.

Social media, however, has taken this stress off the passengers by allowing an instant reach to airlines. In addition, passengers' views can be spread to a wider audience in the online community without being noted by airlines. Such communication could be invisible for business and has the power to affect the brand and passengers' buying behaviour (Finne & Strandvik, 2012; Mangold & Faulds, 2009). The challenges for airlines would be to develop the capacity to establish the existence of the invisible communication and leverage it to their benefits. 'Social activity' ranked third in the list of categories taking 13.8%, while 'Sharing travel experience' was the fifth gaining 7.1%. 'Community support' was the least active category, which only received less than 0.5%. This showed that users in the community were reluctant to provide information or support each other.

Interactive or non-interactive communication

Of the 1875 airline-authored messages, a predominant 91.7% is interactive communication. Among them, only 1% was functional interactivity, which was achieved through the interface's capacity of Facebook enabling a machine–people communication. Nine carriers (45%) attempted functional interactivity on Facebook by inviting users to participate in a poll, quiz, a contest or a vote. Specifically, Swiss International Air Lines, Jet Airways and Mango Airlines contributed 17.6% each; Qatar Airways and Air New Zealand accounted for 11.8% each; and Air Canada, Qantas Airways, easyJet and Kulula Airlines took 5.9% each. Ninety-nine per cent of the interactive communication belonged to contingent interactivity, which was realized through two-way conversations exchanged between airlines and their Facebook users. Eighteen airlines (90%), except Ryanair and Tigerair Singapore, engaged in contingent interactivity. KLM was the most engaging communicator with its Facebook users contributing 38.9% contingent interactive posts followed by Jetstar Airways (12.4%). Table 4 is a summary of interactive communication compared with non-interactive communication, with a breakdown of functional and contingent interactivity.

Types of messages likely to result in functional interactivity

A poll, quiz, a contest or an invitation is most likely to result in functional interactive communication. An airline's quizzes or invitations to vote posted on Facebook are usually related to its operation, its destinations served or aviation technology. For example, Qatar Airways asked its Facebook users to make a guess of one of its destinations served. Jet Airways, instead, challenged its Facebook users with the knowledge of the black box carried by an aircraft, and Kulula Airlines invited its Facebook users to vote for its excellence in social media customer service. Table 5 gives exemplars of airline-authored functional interactive posts. All functional interactive messages were initiated by airlines.

Types of messages likely to result in contingent interactivity

Messages of the nature of 'Information seeking', 'Compliment', 'Grievance' and 'Sharing travel experience' are all likely to result in contingent interactivity as these types of

Table 4. Summary of interactive versus non-interactive posts.

Airlines	Interactive				Non-interactive
	Functional	Percentage	Contingent	Percentage	Percentage
KLM	0	0.0	663	38.9	9	5.8
Swiss International Air Lines	3	17.6	27	1.6	8	5.2
United Airlines	0	0.0	27	1.6	2	1.3
Air Canada	1	5.9	187	11.0	3	1.9
South Africa Airways	0	0.0	4	0.2	4	2.6
EGYPTAIR	0	0.0	94	5.5	31	20.0
Jet Airways	3	17.6	8	0.5	10	6.5
Qatar Airways	2	11.8	2	0.1	10	6.5
Qantas Airways	1	5.9	77	4.5	7	4.5
Air New Zealand	2	11.8	16	0.9	8	5.2
Ryanair	0	0.0	0	0.0	0	0.0
easyJet	1	5.9	80	4.7	1	0.6
Virgin America	0	0.0	46	2.7	8	5.2
Southwest Airlines	0	0.0	32	1.9	16	10.3
Kulula Airlines	1	5.9	68	4.0	2	1.3
Mango Airlines	3	17.6	65	3.8	14	9.0
Tigerair Singapore	0	0.0	0	0.0	3	1.9
AirAsia	0	0.0	83	4.9	11	7.1
Jetstar Airways	0	0.0	212	12.4	3	1.9
Tigerair Australia	0	0.0	12	0.7	5	3.2
Total	17	100.0	1703	100.0	155	100.0

Table 5. Exemplars of functional interactive communication.

Case 1

 Jet Airways

If you are crazy about airplanes, you should know this! What is the colour of the black box?

Case 2

 **kulula**

There is no time to was, we have 3 days left to gather votes for 'Excellence in social media customer service'. We need your vote to show the world that a South African airline is up there with the best in the world.

---

messages either necessitate a response or arouse peer users' curiosity, interest or empathy that invites contribution to and participation in the communication. An exemplar of contingent interactive communication was a request for a follow-up of an outstanding issue in relation to a carrier, a query, a complaint or a compliment about a carrier's service. For instance, a passenger of Swiss International Air Lines posted a follow-up message on its Facebook page asking for updates about the misplaced luggage. The airline replied sympathetically with explicit contact details and procedures for the user to take further action. A KLM user posted a complaint in relation to the change of flight date that resulted in four follow-up message exchanges. All contingent interactive messages were initiated by users. Table 6 shows exemplars of such contingent interactive communication between airlines and their Facebook users.

Non-interactive communication and types of messages likely to result in non-interactivity

Around 8% of airline-authored messages were non-interactive communication. These were messages that failed to generate any responses or follow-ups from Facebook users. This type of message was predominantly airline-authored 'Advertisements, sales and promotion announcements', 'Sponsorship', 'Real-time updates', and 'Social activity' and Facebook user-authored 'Social activity'. Almost all airlines had non-interactive communication messages, except Ryanair. Among them, EGYPTAIR took the biggest percentage (20%) of non-interactivity followed by Southwest Airlines (10.3%) (Table 4).

Differences between FSCs and LCCs and their respective users

Overall, FSCs were more active in generating posts than LCCs, with FSCs contributing 60% and LCCs 40%. A Chi-square test was conducted which confirmed this observation ($\chi^2 = 13.467$, $df = 6$, $n = 172$, $p < .05$, Cramer's $V = 0.28$) (Table 7). Specifically, FSCs were more active in generating five categories of messages which included 'Real-time updates' (100%), 'Miscellaneous' (73%), 'Social activity' (71%), 'Poll/contest' (71%) and 'Sponsorship' (69%), while LCCs were only active in creating one category of message, that is, 'Information provision' (100%). Both FSCs and LCCs were interested

Table 6. Exemplars of contingent interactive communication.

Case 1

Two days without luggage in Muscat and no body of Swiss Air doesn't know where my luggage is. Is it normal????

 Swiss International Air Lines

Dear [user], the handling agent for SWISS lost luggage in Muscat is OMAN AIR S.A.O.C. with phone number 968 24 519 504. Have you tried to give them a call? Also you can purchase necessities, but please keep your receipts and claim via customer service http://bit.ly/o5dnrD once back home and you will be reimbursed for the essentials. I do hope your luggage is found soon. Kind regards, [airline representative]

Case 2

Nice page KLM ….what about the 900€ which I've lost recently because I'd like to postpone my flight with one month? Why is it so impossible to change dates? Only because of rules? My seat will be sold to someone else anyway … that's twice 900€ for KLM … …

 KLM

Hello [user], we can only imagine your frustration about this issue. However, when you purchase a ticket, you agree with its terms and conditions, and so do we. By these is that we abide by and we can't bend them, let alone break them … .

Don't wish me all the best … please … with ONE (oke, maybe 2 …) click of your mouse, just as much trouble as putting your reply on this page, you can change my travel date, and sell my empty seat to another customer. Your words are beautiful and well chosen, but my 900 euro's are gone. Does KLM realise how much money this is for an average traveller with a wife and 2 year old in Brazil?

 KLM

Hello [user], we'd be more than happy to double check your ticket conditions and maybe we're able to clarify this for you. Let us know if you'd like us to contact you via private messages to have a look at your reservation and come back to you with more information.

in creating 'Advertising, sales and promotion' and 'Social activity' posts but less interested in updating information (Figure 1).

This is an interesting note. In the airline industry, FSCs mainly focus on business travellers or those less price-sensitive leisure passengers and compete on service quality with extensive route networks, frequent scheduling and attentive pre-flight and onboard services. They also maintain Frequent Flyer Programs as part of the standard package to retain passenger loyalty (Klophaus, 2005). LCCs, in contrast, target price-sensitive leisure passengers and have adopted a variety of strategies to improve productivity and reduce costs by stripping off all frill services where possible. Passengers are educated to be self-reliant to book tickets, organize their seat allocation, drop bags and arrange connecting flights, where

Table 7. Chi-square test.

Categories	Airline-initiated posts			Categories	User-authored posts		
	FSCs	LCCs	Total		FSC users	LCC users	Total
Advertising, sales and promotion	39 (50%)	39 (50%)	78	Compliment	134 (36%)	235 (64%)	369
Information provision	0 (0%)	3 (100%)	3	Grievance	435 (51%)	419 (49%)	854
Sponsorship	9 (69%)	4 (31%)	13	Sharing travel experience	122 (58%)	89 (42%)	211
Real-time updates	2 (100%)	0 (0%)	2	Information seeking	540 (56%)	423 (44%)	963
Social activity	34 (71%)	14 (29%)	48	Community support	5 (38%)	8 (62%)	13
Poll/contest	12 (71%)	5 (29%)	17	Social activity	246 (59%)	168 (41%)	414
Miscellaneous	8 (73%)	3 (27%)	11	Miscellaneous	106 (64%)	60 (36%)	166
Total	104 (60%)	68 (40%)	172	Total	1588 (53%)	1402 (47%)	2990

Pearson chi-square = 13.467, p-value = .036*
Cramer's V value = 0.28, p-value = .036*

Pearson chi-square = 64.133, p-value = .000***
Cramer's V value = 0.146, p-value = .000***

necessary. Accordingly, FSCs and LCCs employ different marketing communication channels to connect with their targeted segments. For example, FSCs tend to follow their business passengers' tastes and communicate with them via elite media such as CNN, The Economist, The Financial Times and Forbes, while LCCs rely heavily on their individual corporate websites, although popular local newspapers and radio are also utilized. LCCs are more dependent on corporate websites for their marketing and sales activity as e-commerce and a paperless office is one of the best practices adopted to reduce costs. They are more inclined to new technologies and innovation to cost deduction and productivity improvement. However, this research shows that LCCs have lagged behind FSCs in exploiting the prospective functions of social media. While establishing the rationale would be beyond the scope of this research, it certainly deserves further investigation.

Two differences were established between FSC and LCC users. First, FSC users contributed more posts, accounting for 53% compared to LCC users' 47%. This result was also ascertained by a Chi-square test ($\chi^2 = 64.133$, $df = 6$, $n = 2990$, $p < .001$, Cramer's $V = 0.146$) (Table 7). Specifically, FSC users were more active in creating five types of messages which included 'Miscellaneous' (64%), 'Social activity' (59%), 'Sharing travel experience' (58%), 'Information seeking' (56%) and 'Grievance' (51%), while LCC users only outperformed FSC users in 'Compliment' (64%) and 'Community support' (62%). Second, FSC and LCC users demonstrated differences in generating content although the messages appeared to fall into their respective categories. For example, both FSC and LCC users searched for relevant information. However, FSC users were more interested in learning how to organize their trips, such as upgrading baggage allowance while LCC users were keener to find out the special deals and bargains. More than 30% of all LCC users' queries attribute to this type of information seeking, a phenomenon worth LCCs' close attention. It implies that LCCs need to consider how and when to communicate their sales promotions to their targeted market segments via social media and how to turn these users into prospective buyers to optimize ticket sales. This crucial intelligence would allow LCCs to establish what destinations their social media users are interested in flying to and which season of the year they want to book the flight. It would also facilitate LCCs to make informed planning of their route networks, flight scheduling and inventory management.

Discussion

Social media has evolved out of a static web to a dynamic and interactive application with a profound impact on both businesses and consumers. It is argued that social media has the most potential to revolutionize the way businesses operate in every aspect by bringing the customers into play. Our study has ascertained the following. First, airlines have exploited various functions of Facebook, which include: communication channel for marketing activities; distribution channel to optimize sales and service recovery mechanism to recover service failures. These findings confirm other researchers who explored how other industries such as hospitality, banking, education, retail and charities have utilized Facebook.

Next, when using social media as a marketing communication channel, airlines still adopt an inside-out approach in designing and developing messages. One-way advertisement was the primary marketing message posted on social media, which is less inviting for participation. However, what is worth noting is the change of language style of advertisement and sales promotion announcements, which is short, vivid and informal. This is in contrast to what Hvass and Munar (2012) have established that airlines' tone is authoritative. It can be argued that the posts were collected from different airlines in different countries and the timing of data collection varied. Nevertheless, our findings demonstrate that airlines are following the principles of managing social media presence, as suggested by Kaplan and Haenlein (2010), which are active, interesting, humble, informal and honest.

In addition, it is encouraging to observe that airlines have started to embrace an outside-in approach by involving customers into their marketing communication process, resulting in interactivity. They achieve this by deliberately designing and executing engaging messages such as polls, quizzes and contests in relation to their operation and industry technology to attract Facebook users to visit the page, hence stimulating more traffic to realize functional interactivity. What is more fascinating is that airlines have demonstrated their capability of managing contingent interactive communication with their Facebook users. More contingent interactive messages were recorded than functional interactive messages between airlines and their respective Facebook users, showcasing airlines' commitment to entering into a direct conversation with their customers to foster relationships. This achievement is attributable to the convenience and efficiency of the virtual communication forum available on Facebook, whereby an instant interpersonal dialogue is enabled. It is also a reflection of the genuine efforts of airline employees, who are dealing with their Facebook users directly. The intimate, friendly and personalized messages composed by airline staff were able to promote immediacy, personal presence and empathy, which not only enhances users' engagement with the website resulting in increased interactivity but also facilitates relationships building and development (Trammell, Williams, Postelnicu, & Landreville, 2006).

Finally, airlines do not engage in the same level of interactivity on Facebook as shown by the Interactivity Performance Matrix (Figure 3). Six airlines plotted into the Leader's quadrant, which included KLM, Jetstar Airways, Air Canada, AirAsia, Qantas Airways and easyJet. These airlines have demonstrated their capacity to engage in an interactive communication with their Facebook users, which benefits their business–customer relationship building and development. KLM outperformed the rest by not only interactively communicating with their users, but also by their users' overwhelming enthusiasm for interacting with the carrier. KLM alone was responsible for 38.9% of interactive posts, compared with 12.4% of interactive posts from Jetstar. KLM's Facebook users were also the largest contributors to user-generated posts, claiming almost one-fifth (19.6%) of the total.

The Laggard quadrant contained another six airlines, which were Swiss International Air Lines, Tigerair Australia, Jet Airways, Qatar Airways, Tigerair Singapore and Ryanair. These airlines were less active in generating posts on their Facebook pages or in engaging their customers in the communication process. Likewise, their Facebook users did not show much interest in interacting with the airlines. Hence, both parties remained relatively dormant on Facebook. From a business perspective, an interactive communication requires constant attention from the businesses, which may be more than a smaller or a low-margin business can afford to provide. From an individual user's perspective, a user, although having the opportunity to interact with the businesses who then makes an attempt to do so, might not necessarily get an immediate response, hence showing less satisfaction than the one who has no such opportunity at all. These businesses face both internal and external challenges. A thorough auditing is essential to ensure that they have the resources and capacity to embrace social media. They also need to determine how to engage their Facebook users to ensure that they optimize the technology opportunities to reach their customers and establish relationships.

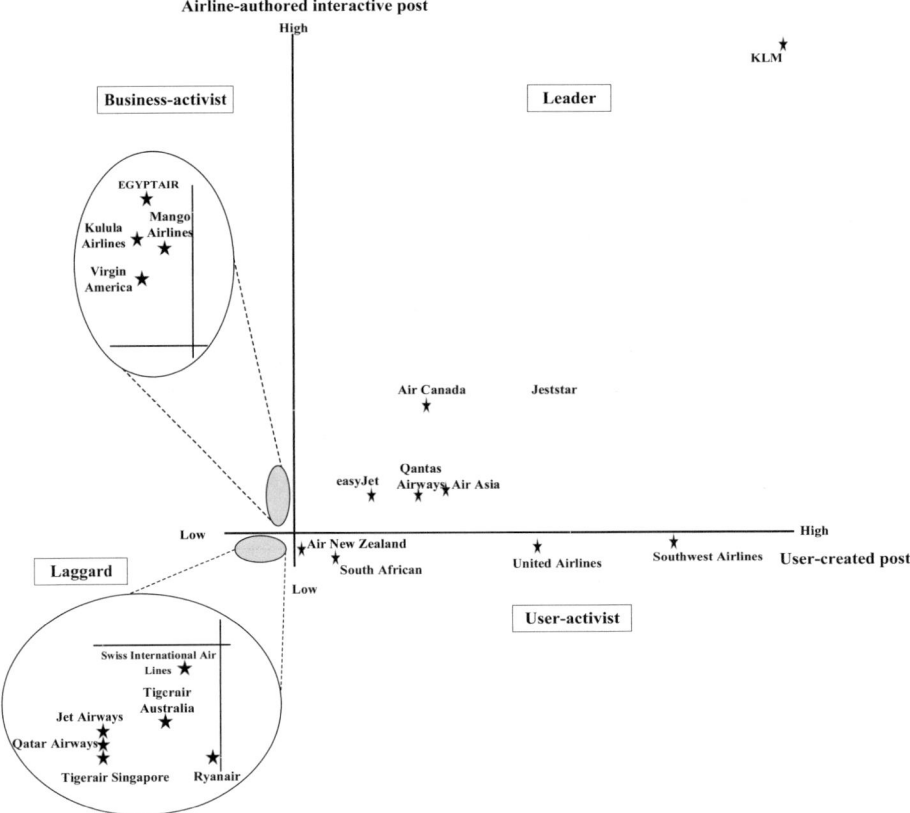

Figure 3. Interactivity Performance Matrix.

The User-activist quadrant collected four carriers, which included Southwest Airlines, United Airlines, South African Airways and Air New Zealand. The users were far more enthusiastic for posting on the carriers' respective Facebook pages though they failed to have received the same level of zest from their host airlines, which resulted in an imbalance of posts. This demonstrates that consumers are increasingly gaining more control in this marketing communication process where the interactivity feature of the media becomes inherent. The inability of businesses to action and react to consumers will open the door for advert e-Word-of-Mouth and viral marketing effect on their brand and reputation. Businesses therefore need to ensure they have an effective social media strategy and are able to implement the strategy. In addition, they need to identify the influential users and make sure these users could become their ambassadors to generate positive effects.

The quadrant of Business-activist gathered another four carriers which included EGYP-TAIR, Kulula Airlines, Mango Airlines and Virgin America. These airlines were keen to post messages on Facebook pages but unfortunately their users did not seem to be stimulated into a conversation with the carriers. A close examination of these carrier-authored posts revealed that a big proportion were non-interactive marketing messages such as advertisements, sales promotions or sponsorship announcements, which failed to attract users' interest to respond. Although there exists a significant correlation between the use of social media and the individual user's personal characteristics, social-demographic backgrounds and computer skills (Amichai-Hamburger & Vinitzky, 2010; Moore & McElroy,

2012), it is nevertheless the capability of businesses that dictate whether they are able to induce users into a conversation on social media (Gulbrandsen & Just, 2011). This remains the most challenging task for any business.

Theoretical contributions

Our contribution to literature is threefold. First and foremost, we streamline and extend the definition and application of functional and contingent interactivity to the most dynamic online environment, for example, social media domain, whereby the interactivity feature is considered inherent. Second, marketing and communication literature recognizes the role of communication and interactivity in relationship marketing but acknowledges the challenges of achieving such interactivity in both an offline and online environment. Our study confirms that communication can become more interactive in social media environment, conducive to relationships formulation and development. Social media has the ability to facilitate both types of interactivity and enhance it to a higher level. Such technology availability lends consumers more power in this interactivity process. Last but not the least, relationship marketing literature asserts that a consumer-centric approach is essential in the marketing communication process in order to create shared meaning of marketing messages and added value resulting in relationship building and development. Our finding extends this stream of academic research by advancing the understanding of consumers' role in this shared meaning creation process. We ascertain that consumers are instrumental in achieving interactive marketing communication and are gaining growing control over communication with the facilitation of social media.

Management implications

Although this research used airlines as the object of the study, our findings have the following management implications applicable to other industries. First of all, businesses need to consider what type of messages they want to create on social media and how to convey such messages. The type of messages is pivotal in fostering interactive communication. Strategically designed and developed messages have the power to arouse users' interest, which encourages their willingness to respond, to be involved and to participate (Gulbrandsen & Just, 2011). Marketing messages such as advertisements and public relation releases that are welcomed in traditional media, such as newspapers and television, are of more an educational style and are less prone to motivate social media users to participate. In contrast, messages such as invitations to a quiz, contest and a vote are more likely to encourage social media users to be involved as they give users an opportunity to apply their knowledge and skills and express themselves.

In developing interactive communication strategies, businesses need to consider how to demonstrate added value to users to participate in online interactive communication. Consumers are reluctant to accept new types of communication with businesses unless they perceive sufficient added value to compensate for the prospective cost incurred as well as for the extra effort invested in mastering and using the technologies (Vlasic & Kesic, 2007). Businesses need to define the added value and determine how it can be delivered. Should personalized communication be considered one of the added values, businesses need a plan and must leverage resources to make it perceivable and attractive.

Furthermore, businesses should endeavour to identify the most influential individuals and engage them to become champions and ambassadors in leading the interactive communication activities. Those highly influential individuals tend not to be easily susceptible.

Once identified and targeted, they should be given incentives to optimize their capacity to influence their peers and lead the community (Aral & Walker, 2012). Finally, businesses should consider what kind of social media strategies they need to have in place and plan what resources are required to implement their social media strategies. Our findings demonstrate that appropriate strategies and their effective implementation are fundamental to ensure a more satisfactory social media communication performance.

Limitations and suggestions for future research

First, a more theoretically vigorous typology of categorizing social media posts needs to be developed and validated. This would help uncover what types of messages are more likely to result in interactivity. A longitudinal study on posts collected from various social media sites is essential to validate the typology of message categorization. Second, further studies are required to establish to what extent the interactive communication on social media will facilitate the relationship building and development. Any findings will shed light on businesses in their understanding of the role of technology, their customers and their relationships with their social media users, which will in turn facilitate the formulation and implementation of corporate strategy. Third, more research is needed to uncover the reasons as to why there exists a big variance in business interactivity performance. Factors such as geographical locations, cultural issues, infrastructure, business models, stage of growth, corporate strategies and their respective segment customers' behaviour need to be examined and analyzed. Finally, further research is necessary to establish why LCCs have lagged behind FSCs in embracing social media. LCCs have been renowned for its innovation and entrepreneurship since they took shape in the late 1970 but are less responsive to social media than FSCs. Any further investigation in this aspect will shed light.

Our research has the following limitations. First, posts were collected in August 2012, which only represented a snapshot of millions of posts on social media. Second, only posts on Facebook were collected and analyzed, which have different features in enabling online communication compared with other social media such as Twitter, YouTube or Linkedin. Future research could be undertaken to include more social media sites.

Acknowledgements

The authors would like to thank the anonymous reviews and the editor in chief of this special issue for their constructive comments. The authors' thanks also go to the five students in Swinburne University of Technology, who are Wesley Robertson, Bo Huang, Jiabei Liu, Ignatius J. Soegeng, and Aaron Yar Lun Wong, for undertaking the tasks of coding and categorizing the themes of posts. Their hard work and commitment is highly appreciated.

Disclosure statement

No potential conflict of interest was reported by the authors.

References

Akehurst, G. (2009). User generated content: The use of blogs for tourism organisations and tourism consumers. *Service Business*, *3*(1), 51–61.
Amichai-Hamburger, Y., & Vinitzky, G. (2010). Social network use and personality. *Computers in Human Behavior*, *26*(6), 1289–1295.

Aral, S., & Walker, D. (2012). Identifying influential and susceptible members of social networks. *Science, 337*(6092), 337–341.

Baird, C. H., & Parasnis, G. (2011). From social media to social customer relationship management. *Strategy & Leadership, 39*(5), 30–37.

Berthon, P. R., Pitt, L. F., Plangger, K., & Shapiro, D. (2012). Marketing meets Web 2.0, social media, and creative consumers: Implications for international marketing strategy. *Business Horizons, 55*(3), 261–271.

Bezjian-Avery, A., Calder, B., & Iacobucci, D. (1998). New media interactive advertising vs. traditional advertising. *Journal of Advertising Research, 38*, 23–32.

Blasco-Arcas, L., Hernandez-Ortega, B. L., & Jimenez-Martinez, J. (2014). Collaborating online: The roles of interactivity and personalization. *The Service Industries Journal, 34*(8), 677–698.

Boczkowski, P. J., & Mitchelstein, E. (2012). How users take advantage of different forms of interactivity on online news sites: Clicking, E-mailing, and commenting. *Human Communication Research, 38*(1), 1–22.

Burton, S., & Soboleva, A. (2011). Interactive or reactive? Marketing with Twitter. *Journal of Consumer Marketing, 28*(7), 491–499.

Carim, L., & Warwick, C. (2013). Use of social media for corporate communications by research-funding organizations in the UK. *Public Relations Review, 39*, 521–525.

Chan, N. L., & Denizci Guillet, B. (2011). Investigation of social media marketing: How does the hotel industry in Hong Kong perform in marketing on social media websites? *Journal of Travel & Tourism Marketing, 28*(4), 345–368.

Cross, R., & Smith, J. (1995). *Customer bonding: Pathway to lasting customer loyalty.* Lincolnwood, IL: NTC Business Book.

Deighton, J., & Grayson, K. (1995). Marketing and seduction: Building exchange relationships by managing social consensus. *Journal of Consumer Research, 21*, 660–676.

Duncan, T., & Moriarty, S. E. (1998). A communication-based marketing model for managing relationships. *Journal of Marketing, 62*(2), 1–13.

Durkin, M., Filbey, L., & McCartan-Quinn, D. (2014). Marketing to the mature learner: Exploring the role of web communications. *The Service Industries Journal, 34*(1), 56–70.

Etter, M. (2013). Reasons for low levels of interactivity: (Non-)interactive CSR communication in Twitter. *Public Relations Review, 39*(5), 606–608.

Etter, M., & Fieseler, C. (2010). On relational capital in social media. *Communication Sciences, 10*(2), 167–189.

Finne, A., & Gronroos, C. (2009). Rethinking marketing communication: From integrated marketing communications to relationship communication. *Journal of Marketing Communications, 15*(2/3), 179–195.

Finne, A., & Strandvik, T. (2012). Invisible communication: A challenge to established marketing communication. *European Business Review, 24*(2), 120–133.

Fred Van Raaij, W. (1998). Interactive communication: Consumer power and initiative. *Journal of Marketing Communications, 4*(1), 1–8.

Goh, D. H., & Lee, C. S. (2011). An analysis of tweets in response to the death of Michael Jackson. *Aslib Proceedings, 63*(5), 432–444.

Gulbrandsen, I. T., & Just, S. N. (2011). The collaborative paradigm: Towards an invitational and participatory concept of online communication. *Media, Culture & Society, 33*(7), 1095–1108.

Hansson, L., Wrangmo, A., & Søilen, K. S. (2013). Optimal ways for companies to use Facebook as a marketing channel. *Journal of Information, Communication and Ethics in Society, 11*(2), 112–126.

Hays, S., Page, S. J., & Buhalis, D. (2013). Social media as a destination marketing tool: Its use by national tourism organizations. *Current Issues in Tourism, 16*(3), 211–239.

Heeter, C. (1989). Implications of new interactive technologies for conceptualizing communication. In J. Savaggio & J. Bryant (Eds.), *Media use in the information age: Emerging patterns of adoption and consumer use* (pp. 217–235). London: Routledge.

Hunt, D., Atkin, D., & Krishnan, A. (2012). The influence of computer-mediated communication apprehension on motives for Facebook use. *Journal of Broadcasting & Electronic Media, 56*(2), 187–202.

Hvass, K. A., & Munar, A. M. (2012). The takeoff of social media in tourism. *Journal of Vacation Marketing, 18*(2), 93–103.

Jansen, B. J., Zhang, M., Sobel, K., & Chowdury, A. (2009). Twitter power: Tweets as electronic word of mouth. *Journal of the American Society for Information Science and Technology, 60*(9), 1–20.

Jenkins-Guarnieri, M. A., Wright, S. L., & Hudiburgh, L. M. (2012). The relationships among attachment style, personality traits, interpersonal competency, and Facebook use. *Journal of Applied Developmental Psychology, 33*(2012), 294–301.

Johnson, G. J., Bruner II, G. C., & Kumar, A. (2006). Interactivity and its facets revisited: Theory and empirical test. *Journal of Advertising, 35*(4), 35–52.

Kaplan, A. M., & Haenlein, M. (2010). Users of the world, unite! The challenges and opportunities of social media. *Business Horizons, 53*(1), 59–68.

Kaplan, A. M., & Haenlein, M. (2011). Two hearts in three-quarter time: How to waltz the social media/viral marketing dance. *Business Horizons, 54*(3), 253–263.

Kasavana, M. L., Nusair, K., & Teodosic, K. (2010). Online social networking: Redefining the human web. *Journal of Hospitality and Tourism Technology, 1*(1), 68–82.

Kelleher, T. (2009). Conversational voice, communicated commitment, and public relations outcomes in interactive online communication. *Journal of Communication, 59*, 172–188.

Klophaus, R. (2005). Frequent flyer programs for European low-cost airlines: Prospects, risks and implementation guidelines. *Journal of Air Transport Management, 11*(5), 348–353.

Kozinets, R. V., de Valck, K., Wojnicki, A. C., & Wilner, S. J. S. (2010). Networked narratives: Understanding word-of-mouth marketing in online communities. *Journal of Marketing, 74*(2), 71–89.

Kunz, M. B., & Hackworth, B. A. (2011). Are consumers following retailers to social networks? *Academy of Marketing Studies Journal, 15*(2), 1–22.

Ledingham, J. A. (2003). Explicating relationship management as a general theory of public relations. *Journal of Public Relations Research, 15*, 181–198.

Lerner, J. S., Han, S., & Keltner, D. (2007). Feelings and consumer decision making: Extending the appraisal-tendency framework. *Journal of Consumer Psychology, 17*(3), 181–187.

Leung, X., & Bai, B. (2013). How motivation, opportunity, and ability impact travellers' social media involvement and revisit intention. *Journal of Travel & Tourism Marketing, 30*(1–2), 58–77.

Li, X., & Wang, Y. C. (2011). China in the eyes of western travelers as represented in travel blogs. *Journal of Travel & Tourism Marketing, 28*(7), 689–719.

Lindberg-Repo, K. (2001). *Customer relationship communication – Analysing communication from a value generating perspective.* Helsinki, Finland: Swedish School of Economics.

Mangold, W. G., & Faulds, D. J. (2009). Social media: The new hybrid element of the promotion mix. *Business Horizons, 52*(4), 357–365.

Merchant, R. M., Elmer, S., & Lurie, N. (2011). Integrating social media into emergency-preparedness efforts. *New England Journal of Medicine, 365*, 289–291.

Mitic, M., & Kapoulas, A. (2012). Understanding the role of social media in bank marketing. *Marketing Intelligence & Planning, 30*(6), 668–686.

Moore, K., & McElroy, J. C. (2012). The influence of personality on Facebook usage, wall postings, and regret. *Computers in Human Behavior, 28*(January), 267–274.

Morgan, R., & Hunt, S. (1994). The commitment-trust theory of relationship marketing. *Journal of Marketing, 58*(3), 20–38.

Pantelidis, I. S. (2010). Electronic meal experience: A content analysis of online restaurant comments. *Cornell Hospitality Quarterly, 51*(4), 483–491.

Rafaeli, S. (1988). Interactivity: From new media to communication. In R. Hawkins, J. Weimann, & S. Pingree (Eds.), *Advancing communication science: Merging mass and interpersonal processes* (pp. 110–134). Newbury Park, CA: Sage.

Schmallegger, D., & Carson, D. (2008). Blogs in tourism: Changing approaches to information exchange. *Journal of Vacation Marketing, 14*(2), 99–110.

Shih, C. (2009). *The Facebook era: Tapping online social networks to build better products, reach new audiences, and sell more stuff.* Boston, MA: Prentice Hall.

Sreenivasan, N. D., Lee, C. S., & Goh, D. H. L. (2012). Tweeting the friendly skies: Investigating information exchange among Twitter users about airlines. *Program: Electronic Library & Information Systems, 46*(1), 21–42.

Sundar, S. S., Kalyanaraman, S., & Brown, J. (2003). Explicating web site interactivity: Impression formation effects in political campaign sites. *Communication Research, 30*, 30–59.

Swain, W. N. (2005). Perception of interactivity and consumer control in marketing communications. *Journal of Interactive Advertising, 6*(1), 82–92.

Thevenot, G. (2007). Blogging as a social media. *Tourism and Hospitality Research, 7*(3–4), 287–289.

Trammell, K. D., Williams, A. P., Postelnicu, M., & Landreville, K. D. (2006). Evolution of online campaigning: Increasing interactivity in candidate web sites and blogs through text and technical features. *Mass Communication & Society, 9*(1), 21–44.

Vlasic, G., & Kesic, T. (2007). Analysis of consumer's attitudes toward interactivity and relationship personalization as contemporary developments in interactive marketing communication. *Journal of Marketing Communications, 13*(2), 109–129.

Vorvoreanu, M. (2009). Perceptions of corporations on Facebook: An analysis of Facebook social norms. *Journal of New Communications Research, 4*(1), 67–86.

Walther, J. B., Gay, G., & Hancock, J. T. (2005). How do communication and technology researchers study the Internet? *Communication and Technology, 55*, 632–657.

Xiang, Z., & Gretzel, U. (2010). Role of social media in online travel information search. *Tourism Management, 31*(2), 179–188.

Yoon, D., Choi, S. M., & Sohn, D. (2008). Building customer relationships in an electronic age: The role of interactivity of E-commerce web sites. *Psychology & Marketing, 25*(7), 602–618.

Zywica, J., & Danowski, J. (2008). The faces of Facebookers: Investigating social enhancement and social compensation hypotheses; predicting Facebook™ and offline popularity from sociability and self-Esteem, and mapping the meanings of popularity with semantic networks. *Journal of Computer-Mediated Communication, 14*(1), 1–34.

Index